The Fourth Mind

WHITLEY STRIEBER

Walker & Collier
PUBLISHERS

The Fourth Mind

by Whitley Strieber

Left Brain Right LLC, 1112 Montana Ave, Suite 142, Santa Monica, CA, 90403

First LBR printing, first edition, 2024

Library of Congress Cataloging-in-Publication Data Strieber, Whitley

The Fourth Mind / by Whitley Strieber ISBN:

Hardcover 979-8-9922620-2-5
Softcover 979-8-9922620-0-1
eBook 979-8-9922620-1-8
Audio 979-8-9922620-3-2

Cover design by Ashley Guillory

Printed in the United States of America

First Edition

"A little bit of knowledge can be a dangerous thing; or it can be a vibrant need giving rise to verdant forests and awakening sleeping giants."

— Chan Thomas,
The Adam and Eve Story

This book is dedicated to someone from far away who died for humankind in January 1996, without whose sacrifice on our behalf we would have little chance of overcoming current challenges to our survival, let alone achieving our real goal, which is to express ourselves into the great cosmos, and join to the chorus of all being, the promise and mystery of the people of Earth.

ACKNOWLEDGMENTS

I cannot begin to acknowledge all the people who have been helpful to me in this endeavor. First, I must thank my reading group, who have brought their brilliance, insights, tolerance and patience to what must have often seemed an impossible task: getting through to me. I want also to thank the enormous community of authors and researchers whose work has illuminated my effort, most especially the unsung and derided UFO community, and in particular Dr. Roger Leir, without whose pioneering research this book could never have been written.

CONTENTS

PREFACE

The Implications are Staggering
By Diana Walsh Pasulka

The Fourth Mind is the contemporary *Communion,* a rare book of insight and vision. It is a necessary contribution to the burgeoning field of UAP studies, and is among the best books on the topic. Just as Edwin A. Abbott's visionary masterpiece *Flatland* expanded our understanding of dimensions in the 19th century, Strieber's *The Fourth Mind* brings this revelation into the modern era. Exploring the lost human capacity for fourth-dimensional cognition, Strieber weaves together historical wisdom, religious traditions, and cutting-edge UAP research to uncover a reality hidden in plain sight.

Ancient traditions spoke of extraordinary abilities—the charisms of Catholicism, the siddhis of Buddhism, and the technological mastery of ancient Egypt—as remnants of a deeper human potential that has since been forgotten. Strieber argues that modern UAP encounters, experiencer testimonies, and crash-retrieval materials suggest that the so-called "visitors" possess an advanced form of consciousness—one that exists beyond the limits of our three-dimensional perception.

Just as *Communion* broke open the modern discussion of UAP presence, *The Fourth Mind* offers a profound new synthesis, mapping a pathway to understanding interdimensional intelligence and rediscovering our own latent capacities—abilities once regarded as miraculous: telepathy, levitation, healing, and more.

The implications are staggering. If we can regain this lost knowledge, we may unlock a new era of human evolution—one that has been waiting for us all along.

Dr. Diana Walsh Pasulka, author of *American Cosmic: UFOs, Religion, and Technology,* and *Encounters: Explores of Non-Human Intelligence*; professor at the University of North Carolina, Wilmington

FOREWORD

Them (and Us)

By Jeffrey J. Kripal

This is not another book. This is *the* book.

This is *the* book because it engages in a most remarkable act of comparison from direct and personal experience to the documents and data that are available right now, right here, particularly on Reddit and about Roswell. Whitley begins with *experience.* He then reads the very latest posts and letters in the mirror of his own long experiences. That is how he knows what is likely accurate in them, and what is likely not.

Basically, Strieber uses his own very extensive and consistent contact with the visitors for nearly four decades to paint a picture of what might be going on in the abduction experience and UFO phenomenon. In the process of this from-experience-to-historical-event comparison, he discusses specifics, down to an observation of their excretion process, a long discussion of their immense eyes and eye-coverings, speculation about their genetics and profound interest in our reproductive fluids, a discourse on a particular vertebra, C-1, the extraction of the spine, memories, and experiences, the time-traveling soul, and the once active and now "lost" powers of the pineal gland. Strieber is careful to hedge honestly and consistently with the subjunctive tenses (that "might" and that "we do not know"), but he is also very serious. He knows.

I have heard much of this before from Whitley Strieber himself, and I do not mean that in any ho-hum sense. I was present for some of the events that he recounts in the pages that follow (including the one involving a split Jeff and the "Oh my God!" experience--it happened, I was there, both of me). It is no accident at all that I concluded one of my books with his work and books, asked to co-write an entire book with him, wrote forewords or an afterword for others, and happily received the Communion letters, which are now professionally housed and preserved at Rice University in our Archives of the Impossible[1]. I am grateful beyond words. I am also convinced that Whitley Strieber is describing, as best he can, what happened and what it might mean for all of us. The fact that our forms of knowledge, including our sciences, have not reached the reality of his experiences is not Strieber's fault. It is simply a sign of the future form of knowledge. It what is to come, to employ one of Whitley's many subtitles.

What strikes me about this particular book, *the* book, are two things: its remarkable compassion and its utter physicality.

---

1 See Jeffrey J. Kripal, "The Third Kind: The Visitor Corpus of Whitley Strieber," in *Mutants and Mystics: Science Fiction, Superhero Comics, and the Paranormal* (Chicago: University of Chicago Press, 2011); Jeffrey J. Kripal, "Foreword: Reading as Mutation," in Whitley Strieber, *Solving the Communion Enigma: What Is to Come* (New York: TarcherPerigree, 2013); Whitley Strieber and Jeffrey J. Kripal, *The Super Natural: Why the Unexplained Is Real* (New York: TarcherPerigree, 2017); and Jeffrey J. Kripal, Afterword: The Me within Thee," in Strieber, *Them*. (Walker & Collier, 2023).

Strieber, for example, expresses the greatest concern, even love, for the visitors, the creatures, who appear to be donating and sacrificing themselves as much as preying on us. The visitors demonstrate consistent compassion for us for our soul-blindness and spiritual stupidity (in the materialist West). Hence even the epigraph to this book is about such a "creature" in Brazil, who was clearly afraid and alone but also somehow capable of so much more than we are, or of what we have forgotten.

Whitley speculates with Graham Hancock that such forgetfulness might have something to do with a great worldwide disaster some 12,000 years ago. It may be then, and precisely because of that fantastic natural disaster that involved endless fires on the Earth (from meteors raining down on us), that our species was heavily traumatized and lost its powers. Still, some human beings remember such things in their myths and folklore. Others actualize them, or prey on them. "Nature broke us. It's that simple. And now there are predators and scavengers here seeking to feed off our helpless greatness, and a few saints, too, trying to help the sleeping beauty awaken" (185).

It is all about scale, how big we can think, how far back we can remember. The "über-consciousness" or super-consciousness may not be benign on a particular scale, like a particular human life, but on another level it might be more than nurturing, even designing (212). Consider the extinction of the dinosaurs. Without their disappearance, probably by a meteor strike, we would not be here. Natural disaster can make possible other evolutionary forms.

If we can expand our memory or scale, it is not so difficult to see evidence that human beings possess these superhuman powers or potentials. This is the history of religions. We can and should reflect critically on our ancestors' religious interpretations. This may not be magic at all. This might be a still unknown physics. This might be an unknown energy or plasma. Certainly, the visitors themselves look a lot like "conscious energy" (188). And Strieber is adamant that, in my own terms quoted below, the supernatural is likely really the super natural, a natural process that we do not yet understand with our science.

Hence what we today call telepathy, levitation, the control of magnetism or gravity, precognition, the experience of subtle or exotic energies, and mind-to-machine technology. The Them is also an Us. We are likely a multiform species, somehow related, something in communion, perhaps through other dimensions that we cannot normally know. Still, it is time to remember Them and this fourth dimension. It is time to remember Us. That is the fundamental teaching of *the* book.

Strieber's response to "the lady" is just as striking. Strieber writes in code, but he is really speaking of a kind of mystico-erotic relationship that goes far, far beyond our ordinary sexual needs and relationships. Such a cosmic love and consequent joy is not exclusive to human relationships and attachments. I knew Anne. She was not jealous. She was a part of this. Them is Us.

In a fascinating move that I hope surprises, Strieber speculates that they may not be as intelligent as we imagine them to be, particularly in some areas. Their genetic project may well be imperfect or botched, and they are likely afraid of us. They are afraid partly for *physical* reasons. We are just too big and violent. No wonder they send expendable conscious machines into the gorilla troop. I would, too.

Strieber comes back again and again to the physical nature of the entities, or the visitors, as he calls them. As I noted above, this physicality is what strikes me so as the main through-line of the present book. He writes of copper in bovine and human blood and the long-studied and still unanswered cattle mutilations, bone and muscular structure, the biology of seeing in colors, chromosomes, autism and the inability to speak, the fundamental forces of physics and their fine-tuning, the presence or absence of genitals, and brain structure. He also writes of the appearance of "big insects" and "richly aware machines" that look like something between biology and robotics, a fusion that may be a result of their space-faring ways in the stars.

Much of this sounds like science fiction. But maybe that is close to what is actually the case. Maybe we are living in a science-fiction-like universe, or universes.

There is also subtle humor here, like the time the beings asked Strieber what they could do to stop him from screaming: "given that I was surrounded by giant insects, not a lot" (22).

After the reader stops giggling, one realizes that these instances cannot be slotted away as dreams, apparitions, or visions, although their appearances might occasion all of these altered states of consciousness. This, Strieber speculates, is a species, designed or not designed (there is likely a bit of both involved), that is contacting select human beings for very specific biological, moral, and spiritual purposes. Nuclear disaster and ecological collapse, in particular, are on their radar, so to speak. And their bodies? These look like diving suits, clothes that they can take on and off. They themselves may be something like coherent energy, and they seem to know it.

Strieber describes contact or disclosure as a hopeful, positive, really cosmic situation (and I cannot help but observe that "disclosure" is a near-perfect translation of "apocalypse," Greek for an "uncovering" or a "taking off the lid"). It may be, as some suggest, "sober," but only to specific worldviews.

What emerges from the experiences and the histories certainly looks polytheistic or indigenous to this reader. Others have said the same, including Strieber with his thoughts on the Sanskrit yogic siddhis, the "old gods," an ancient science of the soul, Gnosticism, Daoism, Jainism, Hinduism, Buddhism, human levitation in Roman Catholicism, the war of the gods, and Gurdjieff's kundalini buffer or "kundabuffer" (we do look cut off from the Earth and its energies). I can only encourage such directions. There *are* resources in the history of religions to speak of the powers, if only we can expand our scope outside the narrow lens of contemporary materialist science and the Christian West. There are ways toward a physics unknown.

None of this is to suggest this will not be a severe challenge to any and all past or present worldviews. The shock of the future remains. Human beings can certainly adjust to contact, but we will have to go beyond much of our religion and science. This is what I call impossible thinking.

Consider what Strieber calls the soul. The soul for such an EBO or Extra-Biological-Organism, Strieber speculates, is like a field extended throughout the cosmos, not an expression of the individual or some kind of epiphenomenon, some belief, personal social ego, or specific form of embodiment. There is also speculation here about the central purpose of the visitors to be some kind of *apotheosis* or deification, which Strieber himself reads as a kind of loss, a stepping outside of time and so outside of meaning and newness.

Here is one of the places I differ with Whitley, as there is so much in the history of religions precisely about just such a human deification and experience of eternity, but very little of it is meaningless or not-new. It is much more about the playfulness and joy of Meister Eckhart and the optimistic cosmic evolution of Teilhard de Chardin, about which Strieber also writes. He is ambivalent here, then. But so what of our differences? That is precisely the conversation we need to have on the page and in the public. And maybe both things--the joy and the meaninglessness--are true at different scales.

The visitors have been here for thousands of years, yes, but since the discovery of nuclear energy and our weaponization of it, the situation has grown far more serious. The situation is potentially catastrophic, apocalyptic, now. There is fear in our response. This is what the body does. There is also basic denial, particularly around the "spiritual" or super natural implications of contact and communion (again, that denial seems especially acute in the materialist West).

The historian of religions can observe that Strieber's emphasis on the fourth dimension goes back to Charles Hinton, Edwin Abbott, and modern geometry in the nineteenth century and the new physics of the twentieth and now twenty-first[2]. Strieber is calling us to live, think, and imagine four-dimensionally, not three-dimensionally, as we have evolved to do. The "fourth-dimensional mind" sleeps within us, as he writes. It can see the stream of linear time outside that line, outside that stream.

---

2 I have all my doctoral students read I. P. Couliano, "A Historian's Kit for the Fourth Dimension," in Couliano, Foreword by Lawrence E. Sullivan, *Out of this World: Otherworldly Journeys from Gilgamesh to Albert Einstein* (Boston: Shambalah, 1991). Ioan goes through Hinton, Abbott, and Einstein proposes that the historian of religions write history with the fourth dimension or, in Whitley's language, with the fourth mind.

Again, such an experience of "eternity" (not a whole lot of time, but no-time, outside of time) is an especially common data-point in the history of religions, and it is eminently positive, healing, "salvific." The experience of precognition, prophecy, divination, and the siddhis, as some of our ancestors called these superhuman powers, are also especially common, even if such powers were not always related to the experience of deification or eternity. Strieber writes of them in the context of the visitors and what human experiencers have witnesses. None of them are impossible. They are all normal, that is to say, common to the body of testimony. This is "real magic" caused by natural forces that we do not yet understand or know. This is the super natural, to quote Strieber (and myself).

Let me also observe this. There is some uncanny relationship between the visitors and the writing process, in particular Strieber's writing in a book like *Communion*, whose cover painting, he writes, was "the first real mechanism of contact" (108). And they do not just appear as an entity or image. The visitors work with Strieber to write his books, including via the implant in his ear that produces lines of text that run by in a field seen by his eye. Strieber uses the implant to write. This is one big reason why these books are so special. *Those books are not just Us. They are also Them.* Here at least, in Whitley Strieber, the book is where we meet as a multiform species. The book *is* contact. Reading *is* mutation.

In the end, the visitors do not appear to be here only to rescue us from ourselves. They are also here to rescue themselves. They are here for communion, for our freedom and innocence in the stream of time but also to form something together, something that has not yet existed in the past but will in the future. That fusion is in part the writing and reading again. That both-and, both Them and Us, is ultimately what this book is all about.

We are back to science fiction. This is all very close to "Arrival" (2016), a fundamentally new story that is not the Cold War invasion narrative that dominates almost every other alien move. The "aliens" or Heptapods are here to get us to think about time differently, as circular, and so develop powers that we already possess: precognitive ones. This is not the British colonialism from Mars of *War of the Worlds* (1898). Nor is it the American nationalism and exceptionalism of "Independence Day" (1996). Nor is it the space opera of "Star Wars" (1977). The latter movie and subsequent franchise are really just a Western in disguise, a kind of European settler colonialism now placed in outer space: shoot'em up in the stars. Only a bit of Daoism can rescue the story: the Force, as they say, has both a dark and a light side, and Yoda is very close to yoga. Maybe that is the narrative's attraction at the end of the day: its combination of Western colonialism and Asian religions.

Still, it is time to tell the story differently. It is time to listen, really listen to Whitley Strieber. It is also time to look into the eyes of the visitors and be changed toward the future, to enter an altered state that is only altered from this mundane blindness in which we are presently sunk. The revised cover painting looks out, inviting us in. Can we look into those eyes? They are open now, exposed, vulnerable, glaring, hypnotic. I hope we can look into them as intensely as they look into us. I hope we can look at the Looker.

Jeffrey J. Kripal
J. Newton Rayzor Professor of Philosophy and Religious Thought
Rice University

INTRODUCTION

DANGER AND PROMISE

The Fourth Mind is divided into two parts. The first is a detailed analysis of the physical bodies of one class of the beings that most of us call aliens, but which I think of as visitors, as they enter and leave our world at will, coming from parts unknown. And, in fact, as we move through the fixed reality of their anatomy, we must never forget that their origin is a mystery. Have they come to us out of the stars, out of time, from some other reality, or are they our hidden earthly neighbors, the children of the caves and the dark? Or are they from a mystery too deep for words?

We won't go down those deceptive byways, but stick to what is before us: the sinew and the skin, the strange, crooked genes and the vast brains. And us, also. Always us. Why have we not their skills, their powers?

Who are they is one question. But there is another in these pages: who are we, and what tragedy has befallen us, and how can we find rescue?

The book offers what I believe to be a reasonably accurate description of their anatomy, their genetics, their brains, the way their minds work and, based on what they have communicated about the purpose and aims of consciousness, why they are here.

The second part of the book is about us. It is about what we were like and what our world was like when we possessed the same powers of mind that we observe in them, why we lost them, and how we may regain them.

We are a species in amnesia about our past and denial about our present. Our vision of our past omits the most important thing about it: a catastrophe that almost drove us to extinction and, by so doing, caused us to lose contact with important strengths of mind that have been dormant within us ever since. For the most part, we pretend the catastrophe never happened. But the truth is that we have lost our past and with it all memory of the way our mind can influence the physical world. At this point in our history, crippled though we are, we are once again face to face with an extremely dangerous upheaval.

Someone knows this and is trying to help us. Thus, the book also explores our relationship with them, what they want to do for us and with us, and our most useful possible responses.

I do not refer here to something that will happen in the future. The catastrophe I am referring to is unfolding right now, and our visitors are present and actively engaged with us on many levels. As of the publication of this book, that is still being denied, but that denial is destined to fade. This is why the book exists—to help the process of acceptance move as smoothly as it can.

There is a message from them, or one, or one group, of enormous importance, but which has not been understood. I propose to spend much of the book exploring it, and what it can mean for us as we seek toward a new and more stable way of life for ourselves.

The great human upheaval that started with the beginning of World War One in 1914 and ended with the use of the atomic bomb in Japan in 1945 signaled an unacknowledged desperation so great that it was driving us, as a species, toward suicide. We still are at that point, lingering on the edge of the cliff.

I want us to live. I want every child born and to be born to have a rich, full life. This is why I do what I do. For us. For rich, good lives.

In 1947, the presence I call the visitors, which had been here in a small way for thousands of years, began to ramp up its activities with an idea of trying to rescue us. So far, its efforts have been met with terror, obsessive secrecy, orchestrated derision, and gunfire. This is because, although we have tamed many species, we are not ourselves tame. We are wild creatures and are reacting toward our rescuers as wild animals do to an intruder.

We can do better, and I think that the place to start is by demythologizing our visitors. We will not only explore their anatomy and thought processes in the first part of the book, but also how working with them intelligently and calmly can benefit everybody. They are not only here to help us. They have wants and needs, too. Together, we have to find a balance.

Right now, our relationship is very fraught. This can change, but it is going to take the ordinary people of the world to do it. We cannot succeed if we insist on placing a level of official authority between the average person and the average visitor. That approach is certain to fail. We must, each of us and all of us, come face to face with this fact. But we cannot do it as we are now, crippled by an amnesia about our past that prevents us from truly knowing ourselves.

Also, we don't understand why we have arrived at the planetary crisis we are in. Suffice to say this much right now: We are not to blame. It is because our populations have grown too fast, and that has nothing to do with any decisions we have made and everything to do with the way nature designed us. Our lack of sexual seasonality, absence of fur, prominent genitals and powerful libidos have combined to make overpopulation a biological and mathematical certainty, and it is population pressure that is causing both our wars and the decline of the planet's ability to support us.

In the next few years, we either advance into a much greater life or we sink into what will likely be a terminal decline, as our planet becomes less and less able to support us in our present numbers.

The Fourth Mind is about advancing into that greater life. My plan is to approach the challenge from two directions. The first is the demystification process. I will do this by offering what I believe to be a reasonably accurate and detailed description of our visitors' physical bodies, brain function, and genetics, then moving on to their understanding of what we think of as the spiritual world.

Why do they possess the powers that they do, and we do not? Is it superior technology or something else?

I will show that there is no magic, only powers of nature that we don't understand, but that we *can* understand—as we must, and the new way of thinking and being that their presence demonstrates and promises to us, shows that we can.

The second section of the book will seek to lift our amnesia by looking deep into our own past to see what caused it.

Our visitors have different brains than we do and seem also to have a different relationship with reality; they can control their bodies' density and appearance in ways we cannot—except that we actually can.

The book is called *The Fourth Mind* because it concerns how to move from where we are now, locked in a three-dimensional mind, to learning—or more accurately, re-learning—the secrets of the fourth-dimensional mind that sleeps within us, but is very much alive in them.

I have learned from my life with them that they reason very differently from us. Instead of gently teaching us or explaining things to us, they are goading us into acting on our own behalf by doing things to us that we cannot tolerate. On a personal level, this is how they teach me. If I make a mistake, they don't gently correct me. Rather, they lash out. They do this for the same reason that we might apply physical punishment to an animal we are training: We cannot speak the animal's language and must therefore communicate with him in simple, blunt ways that cannot be mistaken. If contact spreads, we are going to have to get used to this method. It's not pleasant, but it also isn't evil. Their use of it is a big part of the reason that so many of our "insiders" think that they are demonic. They are not demonic. Right now, this method is all they have except under very special circumstances where, if conditions are right, they can communicate mind-to-mind.

Like theirs, our fourth mind, when it reawakens, will see the world from outside the stream of time. For them, this seemingly godlike vision has been a very mixed bag, and the difficulty it has caused them seems to me to be a great part of their own reason for being here. They have gained something close to absolute knowledge, equal to that of who or what created the universe. This has fatally separated them from the creation, in the sense that they know too much to experience much of anything as new. In discovering how to see the future that is hidden, as it were, in the mind of God, they have destroyed their sense of spontaneity.

If I am correct, this makes them the exact opposite of us. For them, little that happens can ever be surprising. For us, every second, always, is new.

If they are indeed trapped in a prison of knowledge, then we can help them spring the trap, and they can, in turn, help us recover fourth mind. I will discuss, in detail, the kind of balanced relationship it will be necessary to achieve if we are going to manage this.

As matters stand, our visitors are at once trying to help us reach a point where we can become the companions they so desperately need, and also exploiting us. By being so elusive, they enforce the secrecy that surrounds their presence, both so that they can continue to exploit us and also so that we won't be overwhelmed by their essentially perfect knowledge and abandon our own path of discovery in favor of feeding off theirs.

By looking at them clearly and analyzing their actions and abilities accurately, we can transform our relationship with them. Instead of undertakers at the death of mankind, we can make them the midwives to our birth. It's up to us. Our choice. In their own way, they are trying to facilitate our survival. Why, and as to exactly how it works, over the course of the book, I will try to make as clear as I am able.

There is an immediate question: Why can I do things like provide a relatively accurate description of their anatomy?

Since 1988, I have known that there were nonhuman bodies in the possession of the United States Department of Defense.

General Arthur Exon, a family friend who was involved with the Air Materiel Command at Wright Field in July 1947, and was officially assigned to it in 1948, spoke of them privately. In 1964, he was assigned as commander of Wright-Patterson Air Force Base, and as late as 1988 told me that he was still consulting with groups there working on the question. Other witnesses have come forward stating that such bodies exist, as we shall see. As both David Grusch and Lue Elizondo, going public under the Whistleblower Protection Act, have also referenced the existence of these bodies, I feel free to discuss them.

Some information about the anatomy and genetics of these bodies, and the spiritual approach of our visitors, has appeared in the public space. At least some of this information, as I will explain, I have reason to believe, is accurate. As a result, I am able to offer, in the first part of this book, a comparison between my observations and descriptions of the bodies offered by a biologist who claims to have studied them.

As to those observations, I would like to make clear that they have occurred in the context of brief encounters between 1985 and the present. There were also apparent encounters in my childhood, but I do not have detailed enough memories from that period. I have had enough direct observational experience in my adult life, though, to have confidence in the accuracy of the more recent memories.

Nevertheless, because of the secrecy that surrounds the subject, my analysis does not come with conventionally authenticated proof. But there is something specific in a document which was deposited anonymously on the social media site Reddit, and sent to me in 2024, that appears only in two places: in the document itself and in my own personal and private observations of the visitors in life, which I have never made public. In my mind, this fact gives the document significant credibility, and means that what is contained in part one of this book is worth considering as a first attempt to bring this information to light. This is not the only similarity between the document and my own observations. There are many, as I shall discuss in detail. There is also information about their genetics in the document that would lead them to display behaviors that I have observed in life, meaning to me that the genetic information offered is probably accurate as well. The document also contains speculations about their spiritual ideas, which I will compare to some of our own, as well as to some that have become clear from my involvement with them. The author of the document doesn't realize it, but this part of it also reveals their reason for coming here, and why they want us to survive.

From the document, it is obvious that extensive research into these cadavers has been done in secret. I hope that this book advances the time that this material will be allowed to be studied in the public space. But that is not for me to decide, nor, in a sense, for any human agency. As I have suggested in previous books, as our visitors could make themselves more publicly known at any time, the secrecy is controlled by them. I suspect that changes in our own policy cannot happen without their at least tacit agreement. One of my purposes in publishing this book is to provide a foundation of understanding that will make them feel able to reveal themselves more openly.

I want us to reach a point where we can have open contact on a personal basis that is fruitful and meaningful to both sides, and advances mankind toward an eventual goal of becoming a cosmic species, which I define as one that understands reality well enough to traverse space and time in the same way that our visitors do, whatever that is.

We must be careful, though. They have made their own mistakes on their journey of life, and, in the end, I suspect that seeing if we can help them correct some of them is yet another reason they are here. Theirs is a rescue mission, but they are not trying just to rescue us, but also themselves. Wary of revealing what they want from us, they creep into our lives by night and steal it.

There is a better way.

PART ONE

BODIES AND POWERS

CHAPTER ONE

THE SECRECY

As this book is being completed, unusual aircraft are flying over many parts of the world, accompanied by government unease and public concern. For now, suffice to say that, if this is not a state or other local player, then our visitors may be acting either independently or through human surrogates who I would speculate have been in place for some time. They are likely to continue to develop this very asymmetric approach to us over the next months and years, likely expanding it in surprising ways. If they are the architects, it is happening now for two reasons: The first is that the United States began, in the fourth quarter of 2024, to redeploy nuclear weapons in the United Kingdom. Drones appeared over RAF Lakenheath where this deployment was being managed, then began appearing over Northern New Jersey. This sensational story caused people to start looking up all over the world, and right now it is all but impossible to tell where the objects are appearing and where public excitement is causing confused reports.

As of this writing, it is unclear if they belong to a state player such as China or Russia, or have some other origin. As they cannot be affected or controlled by our own government, one thing is clear: whoever they are, dominion here is theirs for the taking. Should they return with weapons, there is presently no evidence that we will have any defense. Whether or not we will need it is a question for another time.

Over the years, the visitors have communicated two great concerns about mankind's future. The first is about the danger of nuclear war and the second is about the possibility that our environment may collapse.

The way they are conducting this display, if it is them, would be consistent with warnings that they have delivered in the past. They made many mysterious appearances and engaged in various interventions at both Soviet and American missile sites during the Cold War. This was to warn us about nuclear war. They have also warned thousands of close encounter witnesses about the danger of environmental collapse. Thus, at the same time that the nuclear deployment was taking place in the UK, in the United States, a president was elected who has said that he doesn't believe at all in climate change and will not only undo environmental protections already in place, but even suppress reporting about the situation on the part of governmental scientific institutions.

Thus, both of their great concerns are being ignored at the same time. If the drones are connected with them, this is why they are acting now. There is also a US program intended to develop drones that will protect the homeland from foreign adversaries, and its testbed is located over northern New Jersey, so we shall see.

In any case, *The Fourth Mind* will provide a number of new and practically useful approaches to understanding what contact with our visitors may mean for us, and how it can unfold in our lives in ways that are beneficial to and fulfill the needs of both sides.

While this book advocates disclosure, it is not intended to be hostile to our visitors or to the governments and defense contractors worldwide who have kept secrets related to their presence and intentions. Our official world was long ago manipulated into a position where it had to keep the visitors' secrets. Now it is frightened, very understandably, to reveal what it has done.

One great bar to official disclosure of the visitors' reality and presence here is that it will make it look as if our defense establishment has been outrageously, even criminally, self-serving. But this is just an illusion. Never, from the beginning, did any level of the human world have a choice to act except as it has, and that includes defense establishments worldwide.

The United States Department of Defense makes certain that nobody outside of a small coterie of individuals whose activities are controlled by security clearances has any definite knowledge of anything to do with the visitors. I do have something that they don't, though, which is a lifetime of contact with them, and, in my adult life, opportunities to usefully observe them.

The official rationale is that knowledge of this unknown presence is so strange and so overwhelming that the public isn't ready. And I wouldn't say that this is entirely wrong. It's easy to say "Of course we can deal with it. We are ready." But given that we don't know what 'it' is, how can we possibly know that? In fact, we cannot, not until we have a firm ground of knowledge about the way their minds work and what they want here.

From the publication of *Communion* on, my aim has been to further just this understanding. I am not personally keeping any secrets. My role is to support disclosure by helping people to understand that mysterious 'it.' *The Fourth Mind* is as far as I am able to go at this time. It is my best effort to make use of all I presently understand, in the service of fruitful contact.

Over the years, numerous defense contractors have also gained a stake in prolonging the secrecy because their efforts to exploit found materials and processes have left them with unstable patents on a substantial number of inventions. For official disclosure to happen, there must also be appropriate protections for these companies built into related legislation in all concerned constituencies, by which I mean internationally. There has to be a blanket amnesty covering both criminal and civil liability, and language that protects the questionable patents.

From Truman in 1947 until the present, all U.S. presidential administrations have kept the secret, including that of Jimmy Carter, who promised, after himself seeing a UFO prior to entering office, that he would find out the truth. During his administration, he revealed nothing, and later, when queried about the matter by UFO researcher Stanton Friedman, he refused to discuss it at all. After the end of the Obama administration, presidential advisor John Podesta tweeted that his greatest regret was that he had been unable to get the president to release the UFO information. The Trump administration did not address it, and the Biden administration has done everything possible to derail increasingly strident congressional demands for more disclosure. Similarly, the governments of the United Kingdom, France, Russia, Brazil and China, all of whom possess significant secrets, remain silent on the subject. What the 2025 Trump administration will do, of course, remains to be seen, but I think that they will certainly be compelled to address it in a new way, if only because it is doing that to us right now.

There is much known about UAPs and some of their occupants within the U.S. defense establishment, but it is concentrated on aspects like materials and biology, not so much on aims and motives and the needs behind them, let alone their relationship to the ineffable and intricately resonant part of us that we call the soul—which I always hesitate to bring up, because doing so causes so many people to go instantly into denial. From my perspective, this is because our ever more complex and aggressive material culture has imposed a kind of anesthesia on us. We can no longer detect this part of our normal humanity. By contrast, our visitors are not only capable of understanding the less dense but coherent aspects of personhood—this is their foundational reality. The bodies we will be discussing are, to them, more like suits of clothes are to us than prisons which can be escaped only at death. Incidentally, I don't think it's particularly useful to assume that, if conscious energy, or soul, is part of the human being, we should also embrace past attempts to understand it. As I discussed in *Jesus: A New Vision*, we need to start anew in if we are to engage usefully with this level of our being. Ironically, Jesus knew and understood this two thousand years ago. He was a bit ahead of his time, it would seem.

From personal experience, I have reason to believe not only that the soul exists but also that it is everybody's primary reality, both ours and theirs. The difference between us is that they are not trapped in their bodies and we are. But do we need to be, and were we always?

In the darkest moments of my darkest nights, I see our visitors as predators at large among an innocent species that has no idea how to defend itself, or even that it is being preyed upon. At other times, I see them as—perhaps unwitting—servants of a higher order, putting us under a kind of pressure that will end in our becoming a cosmic species, no longer bound to the physical world and a single planet, and with a consciousness like theirs, that remains clear and objective beyond the physical realm, and in time will be primarily seated there, just like theirs is. Bodies, for us, are also destined to become like suits of clothes that we enter and leave as needed.

Our visitors adhere to the policy of secrecy while engaging with us in invasive and intimate ways that, as we come to understand them better, are going to compel us to empower our side of the relationship.

We have to stand up to them, but to do that with understanding, not with weapons, which are nothing more than a declaration of our helplessness.

The key to unlocking the door is for us to gain a clearer idea, all on our own, of what must be behind it, which leads to the immediate question, "Why don't they just tell us?"

They hide for a number of reasons. One is to keep us passive. Another, more altruistic, is to protect us from the shock of contact. I think, though, that there is a third, deeper reason, which has to do with the different ways our minds and perceptual systems work. We must never forget that no creature can ever have direct contact with the world, not us and not them. We can see reality only through our perceptual systems, and how we see it depends on how those systems work. A bee, seeing into the ultraviolet, perceives the world very differently from the way we do. When we discuss what kind of vision the visitors' eyes afford them, we will go deeper into this issue.

Given what is going to prove to be a very challenging perceptual mismatch, what is needed to push ahead?

Remember the essential challenge of contact: "a new world if you can take it," communicated to Col. Philip Corso many years ago during an encounter in the New Mexico desert. Can we do it? To succeed, we must bear what we find out about the world and ourselves, and that is likely to be harder than we assume.

On the political side, the legislative branch of the government must continue to exert pressure on whatever administration happens to be in power. But this must be done with the understanding that the consistency of policy over all those administrations means that it's going to be an extraordinary challenge. Simply legislating disclosure requirements will not be sufficient. The United States government classifies between 60 million and 70 million documents a year, and procedures to search the classified database are arcane. More than that, much of the UAP information is in the defense contractor space, where it is legally unreachable and, as I pointed out in *Them*, some of it has not been recorded at all, which means that accessing it will require the personal cooperation of the keepers of the secrets. Some of them, deep within the system, who may be in direct contact with the visitors are, in any case, indirectly under their control, and not going to come forward unless allowed to do so by them.

The process of demystification that I am attempting here will, I hope, help to loosen the ties that bind some of these crucial witnesses.

What is the secret that they are being compelled to keep?

Lue Elizondo has described his feeling about it as "somber." If I am even close to being correct about it, I would agree, but with a slight amendment: "Somber indeed, but acceptance is possible and as the old saying goes, 'The truth will make us free.'"

My obligation is the opposite of that of the people who work behind the curtain of classification: I am here to help people engage in contact, not manage it.

With that in mind, never assume that we are somehow 'less.' Human beings are extraordinarily intelligent, durable and capable.

What must happen for the visitors to relax their own policies of secrecy—and therefore for the involved governments to do the same with their codependent ones—is for the average person to leave fear and mythology behind, and find useful understanding. That's where it begins.

To start this process, we must demystify our visitors. To do that, let's begin by examining their anatomy, genetics, brains, and spiritual beliefs, and see how they compare to our own.

CHAPTER TWO

FAR FROM HOME

They were small and broken and the stench was shocking. Initially, they were thought to be children and the flimsy craft, made of tissue-thin sheets of metal on balsa struts, was thought to be a kite or a balloon. Soon, though, it became evident that these were something other than normal children. The next thought was that they were chimeras created using techniques pioneered by the Nazi monster Dr. Josef Mengele, sent to the United States in a balloon to spy on the 509th Bomb Wing at Roswell, at that time the only operational atomic bomber force on the planet.

Then it was understood, and right in the hangar at Roswell Army Air Force Base where the bodies had been brought: these were not children or human beings at all.

General Exon told me after I published *Communion* in 1987 that "everybody from Truman on down knew that what we had found was not of this world within twenty-four hours of our finding it." He also said that he had held one of the bodies in his arms and "it was like a big insect." General Nathan Twining, according to his son, also said privately that there were bodies delivered Air Materiel along with the debris.

When we were making the *Communion* movie in 1988, Anne and I took a side trip to Roswell to meet with witnesses, some of whom were still alive at that time. One of these was Walter Haut, who had remained in Roswell after his retirement.

As press officer, I thought, there was a good chance he might have seen bodies if they had been brought in before the security curtain came down. We met him at his home and had a fairly extensive conversation with him. He described his work as base press officer, and commented that he had been quite surprised on that day when a lid had suddenly been put on the whole thing. He expressed disappointment that the greatest story of his life had been 'spiked.'

He affirmed that Col. Jesse Marcel's description of the debris that had been found in the desert thirty miles from Roswell was unusual, and said that it was his understanding that it had been flown to the Air Materiel Command at Wright Field. I asked him whether the remains of a kite or weather balloon would have received that treatment and he said, simply, "No." I asked for a description of the debris, but only got the response that Jesse was one of the most honest people he had ever known.

When I got to the issue of the existence of bodies, he became, Anne and I both later agreed, rather evasive. It was clear that he was a man who was uncomfortable telling lies, and we concluded later that he had been trying not to lie but also not to tell the truth. We gathered from this that there had been bodies, and he knew this and had probably seen them, but had been sworn to silence. At one point, Anne said to him, "You can't talk about it, can you?" His reply was something like, "If I couldn't tell you, then I wouldn't." We thought that he was telegraphing to us that he had indeed seen bodies, and in hindsight, it seems that this was correct.

Walter died on December 15, 2005, and two years later, his family confirmed our suspicions. He had indeed seen bodies. In 2007, the family released an affidavit that he had sworn and sealed in 2002 in anticipation of death, leaving instructions that it should be made public after his passing.

The affidavit contains 30 paragraphs, which cover Haut's identification and then provide a narrative of what he saw on the day the debris from the crash was brought into the base. Paragraph 13 states that, when he entered the B-29 hangar designated as Building 84, "I was able to see a couple of bodies under a canvas tarpaulin. Only the heads extended beyond the covering, and I was not able to make out any features. The heads did appear larger than normal and the contour of the canvas over the bodies suggested the size of a 10-year-old child. At a later date in Blanchard's office, he would extend his arm about 4 feet above the floor to indicate the height."

David Grusch, who was from November 2021 until April 2023 a Senior GEOINT Capabilities Requirements Officer with the National Geospatial-Intelligence Agency, stated under oath in a congressional hearing on July 26, 2023 that he was aware that, along with crashed disks, "non-human biologics" were found.

The biologist who authored the document that appeared on Reddit does not say that the bodies he claims to have examined at Fort Detrick in Maryland are the same ones that were collected at Roswell. He states, "From the late 2000s to the mid-2010s, I worked as a molecular biologist for a national security contractor in a program to study Exo-Biospheric-Organisms (EBO)." It continues on to describe in detail some of the genetics and anatomy observed and the functional characteristics the anatomy reveals. It also states that the bodies were about five feet tall, meaning that they might not have been the ones that were recovered at Roswell.

Because some of the characteristics described in the document coincide with some very unusual attributes that I have personally observed in such creatures while they were alive, and which I have not found recorded in any public document except that one, I believe that it is worth consideration, as this part of it must, as far as I am concerned, be accurate.

Not everything in the document agrees with my personal observations. I will discuss those differences as well.

Even if somebody eventually comes forward to claim that they made it all up, that cannot be true, as will shortly be seen. The reason is that too much in the document fits with behaviors that only a few people have observed, and, in the one case, only with my own personal observation.

In *Communion* (pp. 185–189), when I told the story of how the face on the cover of the book had been produced, I explained that, in the days that I was working with the artist who painted it, I was seeing a photographic image of the being in my mind's eye, and it was this image that I described to him. On page 252 I added, "I remember very well the eidetic image. I described the joints of the creature I saw as 'insectlike.'" Only later, in 1988, would I meet General Exon, who saw and held one of the bodies, and perceived it the same way. As we shall see, this will become very important in understanding what they are and how they are organized as a society, and, quite possibly, why they are here.

The document describes the cadavers that were said to be under study at Fort Detrick as being about 150 centimeters tall, or five feet. This would have been the height of the ones that were coming around my cabin in the late 1980s, and a little taller than what Walter Haut remembered.

What first arrested my attention about the document was a description of a function that I personally observed and which, to my knowledge, is mentioned publicly only in the document.

I saw this function in action, I believe, in the fall of 1986. The only person I told about it was Budd Hopkins. He was disgusted and thought I must have been dreaming, so I put it aside. To my knowledge, neither of us ever wrote about it. I had no desire to add something so bizarre to my already difficult narrative. Budd never mentioned it to me again, nor I to him.

I believe that the event I am about to describe took place just before dawn. I had either heard or sensed something that woke me up. I went to the bedroom window and saw three of the gray beings standing under the shadows of a grove of pine saplings about fifty feet from the cabin. Usually, when I saw them and tried to engage with them, they immediately disappeared. Some months before this, one of them apparently couldn't manage it, and as I walked closer it grew frantic, gasping and thrashing away through the brush in a panic.

Given the obvious fear it was expressing, I backed off. In those days, I was trying to get them to calm down in my presence, so I had no desire to attempt to catch it or anything like that. Given what I now know about their physical capabilities and their minds, that would have been a fool's errand anyway.

The effort to escape was so desperate because they knew the stakes: If anything substantive ever appears in public about them, they are going to lose the main tool they have for keeping us helpless, which is our ignorance of them. Disclosure of their presence here is probably up to us, but contact is not. Contact is up to them.

This time, I went downstairs and stole quietly out onto our front porch. They did not disappear. In fact, they didn't look at me, so I went out into the yard. Still no change.

I moved a bit closer. When I was perhaps thirty feet away from them, they did look at me. I stopped. I said nothing, and they made no sound. I forced myself to breathe evenly, but they surely must have picked up on my tension. Still, they neither moved nor stopped what they were doing. I heard no telepathic signals. (By that time, this had happened to me more than once, so I was prepared not to react with surprise to telepathic contact, possibly scaring them off, but it did not happen.)

For all the world, remembering back, it was like stealing closer to birds and being aware that they might fly off at any second.

I could see that they were brushing something off their arms. It appeared to be crystalline sheets of material. As soon as they would brush some off, more seemed to ooze out of their skin. There was a faint odor in the air, acrid, and I thought it might be ammonia. I recall telling Budd that this was how they relieved themselves. This is what disgusted him.

From the document: "There are a lot of pores on the skin, crossing from the epidermis to a gland in the hypodermis. These glands and pores are the terminal part of the excretory-sudoriferous system, which could explain the previously mentioned smell." The document reports that a strong smell of "burnt hair and ammonia" is emitted when the "gray film that covers their skin is removed."

This film is described as a biosynthetic covering that is not protective against heat, but does prevent the passage of liquids. This would also explain why nobody reports the ammonia smell or witnesses the removal of waste that I observed. This material is kept against the body when the covering is on. As I recall, the only smell that I noticed while I was inside their place with them on December 26, 1985, was a mild odor of cinnamon. Therefore, they were wearing something over their skin at that time. The author of the document assumes that the covering is protective, but I don't think that this is necessarily the case. Its more important purpose is probably to contain the excretory material. If so, then this supports the theory that these are designed creatures, not an outcome of evolutionary biology.

The document notes that the skin beneath the covering is "very white and the texture is very regular without any hair." I only saw them wiping the crystalline material off their arms, not their whole bodies, which would suggest that they had removed sleeves, and the rest of the bodies were still inside the covering. I don't remember noticing color of the skin on the arms.

One I did observe on intimate terms about a year later (an event described in *The Super Natural*, pp. 141–142) seemed to me to have pale but not white skin. I can recall at one point thinking that she almost looked like a human woman with a very strange face. Unlike the grays I had seen, she had human-looking legs. There was a sort of desperation in her face. She wanted to be appreciated but she appeared very ugly to me. When I so much as thought how ugly she was, she would turn her face away as if slapped, and I would hear her anguished scream in my mind.

Her appearance left me wondering whether they have created instances of themselves that contain some elements of us, and that this is what I was dealing with when I was abducted. The complex insecurities caused by this coalesced into the story "Pain," which is about a powerful feminine presence that cleanses a man's soul with a sort of healing fire. The story was written as the traumatic amnesia I had experienced due to the abduction was lifting, and I was in the process of recalling what had happened to me. In fact, in the middle of the story, there began to be references to UFO material, the first such in my entire body of work.

This entity—I want to say, simply, 'woman'—also differed from the ones described in the document in that she had genitals. She was, therefore, capable of reproduction and possibly more purely biological than they appear to have been. When we discuss their brains, though, the question of the degree of independence of thought that the ones under study possessed will be explored more fully.

The ones that were autopsied seem to me to be less independently functional than this woman. Maybe they were not really independent entities, possibly some sort of symbiotic extension of a creature like her that functions very differently from the way we do in our world—but not, I would think, from the way we might in the future. It's not hard to see that we might end up creating robotic extensions of ourselves that can function independently, even as they are under our overall control, even linked to us in some way.

There will be robots controlled by artificial intelligence in our world very soon, probably within a few years of the publication of this book. They will not be designed biology, not yet, but that may well happen.

Even given a technological confluence like this, it must always be remembered that we are not dealing here with a mind that works like ours. As can be seen in *The Communion Letters,* their logic is not our logic, and their problem-solving choices are often, by our standards, wildly asymmetrical.

It is therefore not surprising that they have chosen to solve the problem of how to engage with us by creating a semi-artificial version of themselves that can be deployed into dangerous situations and is designed to carry out specific tasks.

We do see, in our own environment, many species that have evolved exquisitely detailed, purpose-specific variants. Leaf-cutter ants, as an example, have seven different functional variants in the same species. These different variants also have different brain structures. They are hardwired to task, in other words. It's worth noticing that the soldier ants have the largest brains and eyes, probably to enable them as much flexibility as possible. So maybe the complex, independent creature that managed my abduction, and that spent time with me and my family and became intimate with me, was farther up in their hierarchy than the ones that the Reddit document describes.

The document also mentions something I may have observed, which is that the huge black eyes are actually coverings. "Like the skin, the eyes are covered with a semi-transparent biosynthetic film that offers the same environmental protection, while providing protection against certain wavelengths and light intensity. When the film is removed, a more traditional eye is revealed. It's about three times larger than a human eye and there are no eyelids."

I cannot say for certain that I have seen this, but it seems very familiar. I can imagine—or perhaps remember—such eyes boring into me with great intensity.

When I had the *Communion* portrait painted, I had never seen one without the eye coverings. I would think that few if any people had, which is why, I suppose, they wanted the covering and not the eye itself to be in the image. In the cover on the new edition of *Communion,* the eyes beneath the covering are subtly visible. On the cover of this book, of course, the probable real eye is rendered as precisely as I have managed to describe it to the artist.

Although I did not know this at the time, I have come to understand that the portrait was intended to cause the people they had contacted to realize that what they had thought was a strange dream that lingered in their minds was, in fact, in some way real. For this reason, when the image appeared, it caused a flood of response.

Many thousands of these letters are now housed in the Archives of the Impossible at Rice University. This archive is the first extensive record of contact between mankind and these others. It puts on display the great complexity of contact, and fights the popular illusion that it is a simple, repetitive process. In fact, it is the most complex collective experience humankind has ever had.

I will now compare the anatomical features described in the Reddit document to the behaviors that I and other close encounter witnesses have observed, and determine as best as possible from this what our visitors' physical capabilities and limitations are.The document describes the feet in this way: "At first glance, the feet consist of just two digits, but necropsy soon determined that each toe was made of two fused digits. The medial toe is marginally longer than the distal toe. The feet are relatively longer and narrower than in a human. Their musculature, however, is vestigial."

We would know that they had come around the house at night, especially in snow or when the soil was softened by rain, because they would leave holes that matched the stride of a smallish person moving at a walking gait. I thought that this must mean that they had hooves, like a goat, but feet as described above would leave similar marks if they walk on their toes. Because of another incident, I believe that this is exactly correct. We would sometimes hear their tapping footfalls on the front porch, and the motion-sensitive lights would turn on. This sound would have to have been made by something moving on hard toes, not the soft bottoms of feet.

In addition to this, when they want to move fast, they rise just above the ground and can go very quickly. My son and I saw this happen early one morning at our cabin, a story which I will expand in the next chapter. As to how this is done, I'm not willing to speculate. As these appear to be, at least in part, designed and probably therefore manufactured creatures, it could be both technological and biological, and dependent on physics that we do not yet understand. As recently as October 6, 2024, I was with someone who witnessed this type of motion, although I did not personally see the being who sped past us while we were walking in a forest at night.

Regarding the hands, the document says that they have four digits but no nails, and that there are circular fingerprints, presumably meaning that the fingertips are somewhat rounded. It also states that "Fingers are proportionally much longer than in humans. Unlike humans, finger musculature is entirely intrinsic to the hand. In other words, the muscles used to move the fingers are not in the forearms but entirely located in the hands."

This is from *Communion*, p. 189: "The hands were very long and tapered when in repose, with three fingers and an opposable thumb. When pressed down, the hands became flat, suggesting that they were more pliable than our hands. On the fingers were short, dark nails of a more clawlike appearance than ours."

I have never observed a hand with no nails. The finger ends I have seen round only when they are pressed against a surface. I have observed the hands in function, and the supple movement of the fingers, like snakes, suggests exactly the sort of mobility that musculature "entirely intrinsic to the hand" would have.

Here is a specific incident of this: Our son had a boy up to the cabin at a time when the one I think of as 'the lady' on the cover of *Communion* was coming around nightly. This boy had injured his mouth by putting his tongue in a light socket when he was younger, and had a scar.

Again and again, during the night, she would come to my bedside and project an image into my mind of smoke pouring out of the side of her own mouth, then try to frighten me by suggesting that my son would be kidnapped if he took a cab over to this other boy's flat, which he did fairly frequently.

Both boys were sleeping in the same room of the cabin, and the guest bed was directly across from the door. The next morning, the boy told me that he had seen snake-like fingers coming around the edge of the door and tapping and scraping on it with claw-like nails. He was terribly frightened and wanted to go home, so I drove him back to Manhattan. He left our life.

The anatomy of the fingers, then, as described in the document, is the same, save for the nails, as what I and others observed at my cabin.

The mouth is described as toothless, and it is concluded in the document that they must only be able to consume liquid nutrition. As the cover of *Communion* reveals, I saw a somewhat more complex mouth on the one that peered at me so intently during the abduction experience, but there were others present with almost vestigial mouths and noses, just as the document describes. Obviously, more than one form is involved, but I think that they share many characteristics.

Prior to proceeding with the anatomical structures found in the endoskeleton, nutritional needs, and how those needs are being fulfilled, I would like to discuss communication. They are telepathically fluent, and when we are with them, so are we. So many close encounter accounts, including my own, mention this effect that I think it's safe to assume that the descriptions reflect something real. But that is not the aspect I wish to address here. Our primary means of communication is vocal. We don't use telepathy except when in the presence of the visitors, when it seems quite natural. It has been asserted and essentially proved by Dr. Diane Hennacy Powell, that many unvoiced people on the autism spectrum do have this ability, so it would seem that it can be present naturally among us, and is not the result of some sort of visitor technology.

It would be helpful if they could communicate with us vocally, but the document states that they have no vocal cords, rather that "vocalization is produced by vibration of the wall membrane at the junction between the two air sacs." This refers to air sacs such as those that birds have, which consist of two sets, anterior and posterior. From the document, it would appear that the visitors have only one set, but this would still give them a considerable boost in oxygen exchange efficiency when compared to our lungs.

When birds inhale, air moves into the lungs through the trachea. Most of it is then pulled into the air sacs, which are connected to the lungs. On exhalation, the air in the posterior air sacs flows into the lungs, then moves into the anterior air sacs.

This system enables a continuous flow of air, which floods the blood with oxygen, giving birds their high energy efficiency and, in the case of the visitors, I would assume, feeding their larger brains and also making possible the bursts of speed that, as we shall see, their musculature allows. There is mention of only two air sacs being present in the cadavers.

As to vocalizations, I've heard five different types of vocal sound coming from the grays, but only three of them were probably produced naturally.

During my abduction experience, a voice kept repeating the phrase "What can we do to help you stop screaming?" but it was obviously artificial. (The answer was, given that I was surrounded by giant insects, not a lot.)

I once heard a phrase uttered, "Have joy," but I do not think that it was produced by any vocal process available to the grays. Without vocal cords or something akin to an avian syrinx, it is not going to be possible for them to produce structured sounds like words except in only a very limited way, if at all. However the words "Have Joy" were generated, I very much doubt that they themselves uttered them.

Still, those words were terribly important, which Anne saw at once. She adopted "Have joy" as the mantra of her life, and sought always to meet the challenge, and even in her dying, did just that.

There are three types of vocalizations that I do think the grays produce.

When, after I had come to realize that the abduction of December 26, 1985 had really happened and did apparently involve nonhuman beings, I began trying to go out into the woods at night with the intention of re-engaging. On a couple of occasions, I heard a horrible sound, an unearthly, protracted shrieking. It had powerful emotional content, so it wasn't a machine. But it also couldn't have been an animal because it lasted too long.

I heard it again when we were living in a brownstone in Brooklyn Heights, when an intense encounter took place there, and I found one of them leaping on my back and crying with great intensity in this way as I lurched along a corridor. I cannot say that I fully understand the why of this, but anyone who says that the grays are robotic is quite wrong.

This sound was continuous, not plosive, and could certainly have been produced by a creature with continuous air exchange such as described above.

I have also heard gasps, both from the grays and from beings who seem human but cannot speak except telepathically. These beings, incidentally, are not ethereal or non-physical in any way. They are real flesh and blood, and, judging from my experience with them, at once quite dramatically difficult to be with and almost entirely helpless in normal society.

Given the proficiency of our visitors with genetic manipulation, I fear that they have created genetic mixes who appear human, but are too different from us to ever become part of our society. In *A New World* (pp. 166–169), I described the experience Anne and I had with one of them who had first appeared in the woods behind our old cabin, then followed us to Texas when we lost the cabin due to the financial setbacks that resulted from all the public attacks I experienced.

I have come to believe that this person—a boy—was my child. From some of his actions, I feel that he had offered himself as a sort of guardian for us, but he was just too strange and ominous, and accompanied in Texas by two men who were even worse, so it was all quite impossible.

The other vocalization I have heard is a cry, but it is nothing like a human or animal cry. On one of the two occasions this was produced, it was the most emotionally compelling utterance of any kind that I have ever heard. But it was not complex in the way a bird's call can be complex. The modulation of avian vocalizations depends in part on the syrinx, which enables birds to vary pitch and also separate notes. The document doesn't mention that the cadavers possessed a syrinx. As emotionally compelling as this utterance was, it did not contain separate notes like birdsong. The tone was exquisitely controlled, though, exactly as would be expected if it was being produced by an organ—a membrane—that was being modulated by a complex and subtle mind.

When Anne and my son was nine, we became aware that the being who had been physically intimate with me was taking an interest in him. I would hear her craft come over the house, making the sound of ball bearings ticking together in something that had been spinning very quickly, but was now slowing down. Then the motion-sensitive lights would turn on and the tapping would come as she walked along the porch. Given the encounters I'd had with her, and what Anne and I both perceived as a complete cultural disconnect, not to mention that our son had been taken right out of the house at one point, we were quite concerned and decided that we wanted him out of the house. We sent him off to summer camp.

A couple of nights after he left, she appeared. I heard her craft come down over the roof and into the yard beside his bedroom. A moment later, there were three cries. I could hear the surprise and disappointment in them, as clear as if she had spoken words. The emotional complexity of these sounds was startling.

But at least he wasn't there to be molested, or whatever it was that she had been doing with him.

A few of days later, though, a call came from the director of the camp explaining that he had suddenly disappeared from the face of the earth in front of a large group of children playing soccer. This made us fear that she might be about to take him from us forever. The breathtaking command of reality that this incident revealed caused us to see that we had to bring him home, and we did. I wish that I could say that he remembered any of this, but he doesn't, and perhaps that's just as well. If Anne were still here, she would certainly remember.

The next and last incident illustrates the astonishing power that these vocalizations can communicate. I described it in *Breakthrough* (p. 18), in the following way: "The instant that my hand touched the doorknob there came three sharp, clear cries from the direction of the meadow beyond the woods. I stopped, turned, looked back. Those remain the most emotionally alive, most heartrending sounds I have ever heard. They were so vibrant with love, with longing, with hurt that I can hardly express their impact. I have since realized that they were also incredibly rich, far richer than music, richer than the most emotional of our voices."

"It felt as if some deep, enormous, and lost part of being was calling me from the other side of the woods. This was absolutely real, absolutely physical. If anybody else had been with me, I have no doubt that they would have heard exactly what I did."

It was preceded by what was either the best decision of my life or the worst. I had been asking in my mind for direct contact. I had envisioned, in great detail, sitting down side by side on a bench in our little woods and conversing. Then, one February morning just before dawn, I was awakened by a long trumpet note coming from the woods. I threw on my robe and slippers and rushed outside. As I was hurrying toward a clearing where I could see a large object hovering and making a clanking sound, I paused, trying to take everything in. At once, I heard a rough and nasty-sounding telepathic voice say, "Come on, come on." It sounded menacing. I thought I might go down there and never return. So, I went back to the cabin.

The vocalizations were so perfect, so outside of experience, and so deeply emotional that they caused a profound change in me, one that lingers to this day. I know that there are hearts greater than ours, and emotions deeper than ours. But I still cannot know what this extraordinary presence actually means. What is the connection between the hard-edged telepathic demand and the soft, deeply saddened physical cry?

One thing is obvious: they did not intend to kidnap me. Had that been their objective, they could have done it. I had to make my own decision, and it was respected.

A few moments after I entered the house, there was a visual communication in the form of an image appearing in my mind's eye. It left me understanding that I was still very much a child in all this, but also that I had taken my first steps. The press for disclosure that is increasing among our elected representatives suggests that we are as a society now in the process of doing the same.

I would now like to turn from the echoes of the soul that one finds in the vocalizations to another area, which will yield some different but also very startling and important information.

It begins with a question about an infamous mystery: cattle mutilations, and the unusual needs of a digestive system that has not followed the same pattern as animal evolution has on Earth.

According to the document, what happens along its tract is that nutrients are absorbed and waste matter excreted, as has been discussed, in such a way that there is no need for the sort of stomach we have.

One might ask what sort of nutrition they need. As there are no teeth and no tongue, according to the document, the only food that they can consume would be liquid.

Which nutrients would need to be in this liquid, then? "It is assumed that, given the high metabolic needs of their brains, this food would have a high carbohydrate concentration. In order to meet other metabolic needs, there must also be a high protein content in the food consumed. These two statements are supported by the type of enzyme secreted by the digestive organ. It is therefore speculated that the food consumed is a sort of broth rich in sugar and protein, which probably also has a high copper content."

If they need a food source with a high copper content, the question becomes whether or not there is one readily available on Earth that seems to be being put to some sort of mysterious use. And as it happens, there is a particular food source that has been mysteriously vandalized for years.

The blood of cattle contains abundant copper, and they have the added advantage of being part of our food supply, too, so they can be taken with the assurance that, while the ranchers will be angry, the destruction of the cattle won't cause outright panic. Human beings have an even higher copper content in their blood, but our visitors are no fools. They generally, but not always, leave us alone and make use, instead, of our own prey species—our livestock.

Largely because of Linda Moulton Howe's tireless efforts over now nearly fifty years, denying the strangeness of these events and the inexplicable way in which they must be accomplished requires an active decision to refuse to consider the evidence.

Ms. Howe first reported on a cattle mutilation case in May 1979. The incident occurred in Elsie, Colorado, where a rancher named Tom Miller found one of his cows dead with strange injuries. The animal had been killed by what amounted to a surgery accomplished in an impossible way, using an unknown form of surgical instrument that made unusually precise incisions.

Linda reported that there was no blood found around the carcass, despite the extensive nature of the mutilation. This lack of blood is a common but not universal feature of cattle mutilations. Bovine blood is rich in copper, so I think that it's reasonable to speculate that it may be the basis of a nutritional drink that these entities are using for their sustenance. The liver, also often found to be missing, has a high copper content.

According to the document, "the musculoskeletal system is very ordinary, albeit underdeveloped. Most of the human skeletal muscles have an equivalent." As to the bones, they were reported to contain not marrow but collagen, hydroxyapatite and copper oxide crystals. (In chapter 7, when I discuss exotic means of movement that go far beyond what we now conceive of as propulsion systems, the reason for the presence of copper oxide in the bones will become clearer.)

The collagen would strengthen the bone and the hydroxy-apatite would aid in ion exchange and perform other supportive functions. There are traces of copper oxide in the bones of various earthly creatures, including humans, but the large amount reportedly found in the cadavers would suggest a more important function than it has among creatures that have evolved here.

The document does not state whether the copper is crystalline or loosely packed, but if it is not crystalline or otherwise fused into something more solid like marrow, it would seem that the bones might lack internal stability, and perhaps this is another reason that the creatures are so wary of us. They're fragile.

I do think that it's possible that the copper may function in some esoteric way, perhaps even related to signaling. But its presence makes it clear why they need so much copper in their nutrient mix. It is another sign that the author of the document is correct, and these creatures are designed organisms. This therefore means that things like the copper in their bones is not an adaptation to some unknown environment but the result of a design decision. The copper is there because it is required so that they can carry out needed functions.

It is an electrical conductor and can be an antenna, but the real reason for it may be more esoteric.

The craft are known to generate powerful magnetic fields, and while copper is nonmagnetic, there are a number of ways that it can interact with such fields. When copper is exposed to an active electromagnetic field, eddy currents are created in the copper which generate small magnetic fields of their own. This effect, called Faraday's law of induction, induces repulsion, which could be used in braking the magnetic field, electromagnetic shielding and heat generation. If the objects penetrate into outer space, this last effect might be important to the occupants' survival.

There are so many possible ways that a copper-rich body might interact with a magnetic field that it seems to me likely that this is the reason that the copper is present. If so, then these pilots can be seen to be not just engaging in guidance like our pilots, but as actually being in a state of symbiosis with the craft, literally part of the machine, but probably also able to detach from it and engage in other activities.

I say this because these frail and seemingly robotic creatures also have huge brains, 20 percent larger than ours and containing four lobes instead of three.

Does this mean that they're more intelligent, though?

Not necessarily. But it does mean that they are going to be very, very able when it comes to achieving their mission objectives.

As to what these might be, I think that the leading candidate must be the abductions.

I say this because of the many descriptions of the approach of these creatures to their human objects of interest. The creatures look exactly like what is being described in the document. As an example of how these missions might look to a human who is facing them in the night, I'd like to draw from *The Communion Letters*, as quoted in *Them*.

It was after midnight, and the author of this letter had just been awakened by a strange jingling sound coming from her back yard. She got up and looked out the back window. There, hovering just above the lawn, was a flying saucer.

"I would estimate that they were probably the size of a five- or six-year-old. They were wearing body-hugging dull silver-gray suits that seemed to cover them completely from head to toe ... I saw the fingers of the first to reach my window as they reached up to pull themselves up by the ledge. These were not people fingers. There were only four of them and they were a different color. They seemed to have a bulbous look to them." (Note the "circular fingerprints" description in the document.)

Our correspondent had been a teenager when the event took place, sleeping in a room with her sisters. As is typical of these experiences, nobody else remembered a thing. But why not? The memories could have been suppressed, of course, but there are also possible reasons for this having to do with the way the human brain handles memory, which will be discussed in a later chapter.

If the memories were suppressed, then I would think that the ability has to do with that brain of theirs, the two additional hemispheres that are the seat of this additional level of mind, and manage its exceptional powers.

CHAPTER THREE

BODIES FROM THE BEYOND

The Reddit document is sufficiently detailed to enable me to use quite a number of comparisons from my own life observations to create descriptions of the anatomy of the bodies. It also offers a picture that is anatomically viable, that is to say, the structures it describes would work in real life and in a world like ours.

For example, the document says this about the respiratory system: "Their cellular respiration is equivalent to ours, i.e., they need to oxidize organic components to produce energy." They will therefore need an atmosphere that contains oxygen if they are going to stay alive. That the amount of oxygen in our atmosphere is optimal for them can be inferred from the way they are observed to function. They show no sign of struggle for air, and can engage in high-intensity physical activity. Indeed, they can do better than we can, in the sense that they can move with lightning speed, which I have seen them do on a number of occasions. Were their oxygen uptake less than optimal, they would struggle.

While they can sometimes seem capable of almost instantaneous motion, they can also achieve a level of stillness that is almost uncanny. In fact, when they disappear before our eyes, I wouldn't be surprised to find that this is because they become so still that our brains, which have been processing them as being in motion, come to the conclusion that they are no longer present.

By recording that their muscle mass is primarily type one over type two, the document probably explains these observed abilities. "It should be noted that the proportion of type 1 and type 2 muscle fibers is different from that in a human. Indeed, type 1 outnumbers type 2 by about a factor of 10."

Type 1 muscle, also known as slow-twitch muscle, is found in greater proportion in animals that engage in long endurance running like horses, or snakes, which slowly but steadily constrict their prey. But a 10:1 ratio of slow- to fast-twitch muscle mass implies activities that are more unusual than those of horses and snakes. The bowhead whale, which swims for longer periods than any other cetacean, has a nearly 10:1 ratio of type one to type two muscle, the only animal on record that has a ratio similar to that the gray visitors are said in the document to possess. Bats, which must hang motionless for ten or more hours a day, have a 65 percent to 35 percent ratio of slow to fast-twitch muscle. As they must also engage in periods of rapid flight, in some species as much as six consecutive hours without a rest, they have a lower ratio of type one to type two than the bowhead.

What the ratio present in the gray beings suggests is that they spend long periods of time engaged in some persistent activity that requires steady but not intense muscle pressure, and enables a level of stillness that is not seen in earthly animals. But they must also move in quick bursts. I have seen the use of both muscle groups.

I can offer an observation that probably illustrates one reason that they need this extreme mix of muscle types.

In 1988, when a documentary filmmaker was at our cabin, as I reported in *Breakthrough*, he was awakened by a small man with a huge head staring down at him. When he reacted with fear, the head turned into that of a falcon, and then the being disappeared. What I think had happened was that the creature had reacted to his fear response with a defensive mechanism something like that of a blowfish, in the sense that it had made itself seem more dangerous than it probably was. At present, it is not possible to know if this was something done physically, or if it was perhaps caused by some hypnotic process of which we know nothing.

It then disappeared from sight. However, as it turned out, it had not left the house, and was probably still standing a few feet from the living room fold-out couch where the filmmaker had been sleeping. It would have been taking advantage of a combination of intense concentration and its high level of slow-twitch muscles to remain extremely still, and perhaps some light-bending technology to make itself invisible.

As dawn broke, my son and I arrived on a small hill overlooking the cabin deck. (We had been sleeping out because the house was full.) We saw a translucent silver figure go racing across the deck, then dart off among the trees, dodging around them at blinding speed. Meanwhile, the filmmaker and his wife experienced a burst of heat so intense that they leaped up, thinking that the bed was on fire.

To make its quick escape, it apparently had to reduce its invisibility. This caused it to vent heat from the device that was bending light around it. This is why what we saw was translucent. Also, it did not appear to be touching the ground, not quite, which might explain why the muscles of the feet are described in the document as "vestigial." They are not used during intense activity like this.

The slow-twitch muscles, with their slower release of energy, enabled it to remain motionless in our living room. The fast-twitch muscles then enabled the blindingly fast movement my son and I observed.

It should not be forgotten that the extremely low proportion of fast-twitch muscles means that they can only move like that for a short period. The length of this period, and the amount of recovery time they require, can probably be estimated from an examination of the cadavers, and this should certainly be done.

Many predators evolved to function in this way—a slow and careful stalk, followed by a lightning-quick capture of the unsuspecting prey.

This combination of an ability to move with almost instantaneous rapidity, and then to engage in long periods of sustained, low-level muscle pressure, would be very useful in an abduction scenario. First, there is the slow, stealthy approach, then the almost instantaneous capture, then these very small creatures must carry their much larger victim some distance away.

This and other adaptations we shall be discussing are so perfectly suited to task that the question must be asked: Are these biological entities at all, or are they something that has been fabricated? As we explore their anatomy further, I think that the answer will become clear.

In addition to the strange muscle-mass ratio, the eyes appear to be perfectly suited to task. Let's return to what the document says about them. Not only do they seemingly wear eye coverings, as discussed previously, to control the wavelengths and intensity of the light entering their eyes, but "The size of their eyes suggests they have excellent night vision. It seems paradoxical to cover them with a semi-opaque film. Perhaps they only need to wear it in a bright environment. Their sclera is the same color as their skin, the iris is pale gray, and the pupil is black and oversized. The lens is rounder than a human, and the musculature used to adjust focus is more developed. On the retina, there are at least 6 types of cone cells. The responsiveness of each of these 6 types of cones is specific to a wavelength band, with a minimum of overlap between each other. The result is a broader visible spectrum." This means that they can see more of the world than we can and gives them a definite edge over us when it comes to understanding what is happening around them.

The document does not detail which cones do what, or mention rods, which could provide an idea of the sensitivity of night vision but the high level of musculature development around the eyes could mean the same as it does earthly night predators.

The information provided, combined with an awareness of their habits and the animals in our world with which they most consistently identify themselves, allows some educated guesses about how they might see.

First, it is doubtful, to my mind, that they have the same sort of rich color vision that we do, and therefore that their larger number of cone cells might have to do, instead, with extreme night vision. One reason I suggest this is that color vision is not common among nocturnal predators, which is what their observed behavior and anatomy suggest that they are. This is because it is not needed, and if these are designed creatures, then it seems likely that they are not going to be provided with unneeded attributes.

People who have been with the visitors in their spaces do not generally report seeing anything that is notably colorful. I have seen gray, brown and an iridescent blue. Cones are associated with color vision and, in the case of some predators, ultraviolet vision. Rods are what enable night vision, but only in gray scale.

Our own rich color sense evolved long before we were fully human, when our ancestors lived in the jungle, where the greater the ability of an animal to distinguish color, the more successful it would be eluding predators (generally well camouflaged) and at finding food.

How, then, do they see, and what do they see?

Human eyes have three types of cones. Red cones are sensitive to long-wavelength light from yellow to red. Green cones are sensitive to greens and make it possible to distinguish green from red. Blue cones enable us to see blue and distinguish it from yellow.

Our color perception is very complex, and relies not only on the various parts of the spectrum that the cones are sensitive to, but also on the way our brain is designed to process their input.

In my experience, the two forms of our visitors that I have had the most exposure to come in two colors, gray and dark, iridescent blue. The dark blue ones are very different in appearance and function from the gray ones, and will not be discussed in this survey of anatomy, as I have no data about them to call on.

There are a number of reports of the presence of a green-reflecting tapetum lucidum, or reflective coating, at the back of the gray's eyes. One example appears in the account of Corina Saebels in her book *The Collectors*. She found herself confronting some grays in the company of a friend on a roadside in British Columbia in 2002. As they came out of the woods beside the road, the first thing she saw was green, glowing eyes. There are other such stories, so it seems likely that these night hunters do possess tapetum lucidum. If so, then its combination with the size of their eyes, which would also allow room for a substantial population of rods, would mean superb night vision.

The extreme bursts of speed that their muscles allow require eyes that can follow at high speed. And indeed, the document says that the "musculature used to adjust focus is more developed." This is exactly what we would see in a creature that also has the ability to execute ultra-rapid movements. Once again, what is being described in the document supports observations in the field.

But what about the six different types of cones? There is only one creature on Earth that has six different types of cones in its eyes, the mantis shrimp. Even though the visual needs of this creature are obviously far removed from those of the grays, it's worth discussing why they are needed by this animal, because then we can co-relate them with the known habits of the grays and speculate to some degree about what the six different kinds of cones in their eyes may be.

The mantis shrimp has two types of ultraviolet receptors. They may assist in communications, as the creatures have UV-reflective patterns on their bodies. Other animals that possess UV-sensitive cones, such as owls, use them for detecting rodents, whose fur reflects light in this frequency.

Would there be any need for the grays, then, to have cones that are UV sensitive? The UV reflectivity of human skin varies according to the amount of melanin present, so someone with UV-sensitive cones would be able to tell quite a bit about race, gender, and the state of health of a subject of interest if they could see in the ultraviolet spectrum. For example, Vitamin D levels can be measured by analyzing the UV reflectivity of the skin, as can its age. Hair reflects more or less UV light according to how light-colored it is.

Adding all of these factors up, one cone array is probably UV sensitive. As to the possibility that another UV-sensitive receptor can observe identifying marks, a phenomenon is reported in Dr. Roger Leir's book *The Aliens and the Scalpel* that suggests that this might be the case. It is that "a brilliant green fluorescence" can be seen under ultraviolet light on the bodies of some close encounter witnesses. This tends to persist even after bathing, although there is no data to indicate that it might be permanent. Also, fluorescent handprints have been observed in the dwellings of some people claiming close encounters. Unfortunately, the general opprobrium in which the subject is held means that no extensive research has been done into this phenomenon, including no testing of the content of whatever this fluorescent substance may be. So, a second type of UV-sensitive cone is a possibility, although there is no way to tell whether more than one would be needed.

Another type of receptor that is found in some animals for which the greys might have a use is the far-red receptor. These receptors detect longer wavelengths of light in the red part of the spectrum. This could be either a specialized receptor of some sort or, as I think more likely, a cone. The reason that this might be one of the six types of cones they possess is that the ability to see beyond the spectrum of visible light would assist them in finding us, and it does seem that their configuration is in part designed for this purpose.

The human body emits heat in the 8,000 to 15,000 nanometer range, and an eye sensitive to this range in the use of a well-informed mind is going to be able to tell many things about a person being observed in the infrared, and these are going to be critical in an abduction scenario. For example, by watching the speed and intensity of the breath, an assessment of the target's state of awareness and level of apprehension can be made. Thus, I think that infrared vision would be an essential tool for the grays.

To sum up, then, we have a creature with a number of abilities that we do not possess, and which would seem to make it particularly suited to find, evaluate and control human beings. Of course, as we know nothing of the universe, it could be that these same capabilities are more generally useful as well.

If my analysis is correct, then some of these abilities are as follows:

—Muscles that enable both long-duration performance and sudden movement—movement so fast than we can't see it, let alone escape from it.

— Hands that are capable of movements more complex than our own, and which may have a richer complement of nerves in their broader finger pads.

—Eyes that support the ability to move very quickly, and which have the ability to see light frequencies that enable them to evaluate such qualities as our age and state of health and our sex, and to determine our level of awareness.

—An ability to become still in such a way that we can't see them. I have observed them moving with the movements of my own eyes, then suddenly disappearing. I would think that this is related to visual acuity, motor skills and an ability to concentrate that is greater than ours.

—An ability to obtain food from the same food sources that we use.

—An ability to use the magnetic fields around their ships to generate body heat due to the presence of copper in their bones.

Their limited vocalization skill, on the other hand, is less useful, seriously limiting their ability to communicate with us.

If this is a predator, and there is almost certainly some of that in these people, it is not about our bodies, but something deeper within us that we possess and, I would assume, they do not. In a later chapter, we will explore this thoroughly. I might add parenthetically that I use the term 'people' at times rather than, say 'beings' because I see them as individuals. To me, they are discreet individuals with both a communal presence and self-identification, not a faceless hive.

I think that we may be in possession of something that they don't have at all, but which is much wanted, or even necessary to have a life worth living. I suggest this because we are natural biology and they appear to be what I might describe as richly aware machines.

I have suggested this in previous books, but will be able to explore it more deeply when I discuss the material that appears in the Reddit document about their spiritual beliefs, which has significantly clarified the situation.

CHAPTER FOUR

THE KINDNESS OF STRANGERS

This is the story of a gift, given to us at the cost of an unknown life from a distant place. This gift could be the basis for a recovery, over the next few decades, of a lost way of being human, which has the potential to revise the fundamental meaning of our lives, and grant us a far better chance of long-term survival than we presently have any reason to expect.

It is one of the few clear demonstrations of fourth mind in action to appear in modern times.

To understand it, we have to first understand some things about a country with arguably the most powerful and provocative history of UFO encounters in the world. That country is Brazil.

The number of cases in that country is so extensive that even a condensed survey would easily fill an entire book, so what I would like to do is to highlight some typical cases, and then concentrate on events that took place in the small city of Varginha in 1976.

On the evening of August 9 of that year, Cicilio Higinio Pereira was walking home with some friends when they saw a light in the sky that began to come closer. Soon, it seemed to them that it was following them, and they began to run. Cicilio could not keep up with the others and soon found himself almost pinned beneath a strange flying machine. It was so close to him that he could reach up and touch it. He felt a cold wind and heard a faint humming sound. There was a smell of sulfur in the air, and he could see three men, three to four feet tall, inside the object, which was translucent.

The encounter didn't last long, but it left him with serious health issues. Bob Pratt, in his 1996 book UFO Danger Zone, describes Cicilio's situation in this way: "Almost immediately, he felt sick and began throwing up, spitting up bile and suffering mostly from the dry heaves. He ran the rest of the way home as fast as he could, arriving exhausted, confused, scared and still nauseated. He continued to throw up the rest of the night. It was the beginning of a sickness that eventually ended in his death." (Pratt, UFO Danger Zone, p. 195.)

Throughout the 1970s, while abductions were happening in substantial numbers in the United States, attacks like this were taking place across Brazil. The two situations differed in one important respect: the abduction cases being reported in the U.S., Canada, the U.K., Australia, and to some extent in other countries were carefully organized and generally involved the removal of the target individual from their home in the middle of the night, then their return sometime later, often with only a confused memory and little physical harm. Deaths and disappearances were not reported.

The situation in Brazil was different in one critical respect: People were not being abducted while asleep, but rather being attacked while awake, usually while outdoors. They were often injured and certainly always terrified. I have been unable to find any evidence for classic abductions with the removal of sexual material.

What was happening in Brazil seems less organized than the operations to the north. It's as if people were being carefully chosen in the north, but simply hunted down after being spotted at random in the south.

Often, victims were injured by beams of light from above or simply by getting too close to the objects involved. A number of them felt that the beams of light were extracting blood from their bodies, but there is no way to follow this up.

A good example of this is the case, as reported by Pratt, of Claudiomira Rodrigues, who was sleeping in a hammock outside her house in the rural community of Colares, where numerous UFO-generated assaults took place in the late 1970s. There were so many cases that the Brazilian Air Force conducted an extensive investigation in 1977 and 1978. Between two hundred and three hundred people were interviewed.

Claudiomira's experience illustrates the difference between what was happening there and in the Northern Hemisphere. Interestingly, there is almost no evidence of similar incursions taking place in Asia, but this could be because of a lack of communications infrastructure in rural areas, and political suppression of such stories in communist nations.

Claudiomira saw a man standing in an object that was floating nearby. She described his eyes as "very small," in other words not like the distinctive eyes of the grays, which are anything but small. He wore what looked to her like a diving suit, so she really couldn't tell what was there, only that it had a head and two arms. He carried what looked to her like a pistol, which he pointed at her. Three narrow beams of light, like lasers, struck her in the chest. She reported that "It was very hot. I think each time he took blood."

The burns left three pinpoint scars in a triangle on the upper right side of her chest, just above her right breast, meaning that the attack was carried out with great precision. She immediately became very thirsty and found that she could not move her legs. Of course, she was terrified, but there was nothing she could do to prevent what was happening.

She began to scream, which woke up her cousin Maria, sleeping in the next room. Maria began screaming, too. The man and the light disappeared and, a few minutes later, Maria helped Claudiomira get to the hospital, where her burns were treated.

Many of these incidents were quite destructive of the lives of the victims. An example is one that took place in a tiny settlement called Tapiapanema, in the interior of Mosqueiro Island. A seventeen-year-old girl, Silvia Maria Trinidade, who was five months pregnant, was hit by a beam of light from a UFO. She was so distraught that she lost her baby and her marriage failed.

This incident took place on the evening of October 29, 1977, as Silvia and her husband, Benedito, twenty-four, were lying down. "Silvia woke up and saw a light in the sky." The next thing she knew, a beam of light was shining into the house. When it touched her arm, her screams woke up Benedito.

The neighbors all ran outside (the settlement contained only a few houses) and saw the bright object. One of them shot at it with a rifle, and it disappeared.

Silvia was taken down the river to the hospital in Mosqueiro. While she and her husband were in their boat (it was an hour's trip) the UFO reappeared and shone a light down on the river for about fifteen minutes.

Silvia was not burned, but there was a bruise where the light had touched her left arm. She was in the hospital for a week and lost the baby two months later.

Even though the attacks were quite different from the abductions that were taking place to the north, the disregard for the personal integrity of the victims is a universal characteristic.

It is in this context, then, that another case unfolded, this one entirely different from both the attacks and the abductions. It happened some years after the attacks, which subsided across the early eighties.

On January 20, 1996, a series of unusual events unfolded in the vicinity of Varginha, Brazil. They are described detail in Dr. Roger Leir's book *UFO Crash in Brazil*. After summarizing them, I will concentrate on just one aspect, which consists of an interview that Dr. Leir conducted with a group of doctors, collectively referred to as "MP," who carried out an orthopedic surgery to set the broken leg of a nonhuman being who had been brought to their hospital by military personnel. Then, afterwards—the miracle. It is not one that we should assume is impossible to replicate. On the contrary, the ability to perform it, and more besides, is part of our birthright as human beings.

Roger Leir was like so many UFO researchers, unsung, ignored except in that little community, and absolutely determined. Certain people, when they find something important hidden beneath a lie, will not stop trying to expose the truth. Roger Leir was such a person.

He suffered for it, of course, as we all do. Like John Mack, his license to practice was challenged. He had to deal with a nuisance suit. His practice declined.

He did not stop, and we are all the richer for it.

I knew Dr. Leir well, and when he first heard of the case in 2002, he called me and asked me to go to Brazil with him. My schedule, unfortunately, did not permit it. In retrospect, I should have dropped everything. When he was in Varginha, he called me again to tell me that he had just conducted what he thought was one of most amazing interviews ever done in the history of UFO research. Actually, it was more than that. It stands as one of the most amazing interviews ever done anywhere, on any subject.

What had happened that led up to it has proved to be one of the most extraordinary UFO cases ever recorded, joined in that regard by the Ariel School incident that Dr. John Mack investigated in Zimbabwe.

The overall case is very complex, and the usual false denials can be found in abundance online, but I never knew Dr. Leir to lie, and he does report in his book that, when he visited Robert Bigelow at his offices in Las Vegas and told this story to him and a group of scientists who were with him, Mr. Bigelow was convinced that he was telling the truth. I was, too, when he told me. There was never the slightest doubt in my mind about his truthfulness. But is the story accurate?

Surprisingly, on July 12, 1996, the *Wall Street Journal* published a story about the events of the previous January in Varginha. From the story: "On a Saturday afternoon stroll in January, a trio of young women decided to take a shortcut home through a vacant lot. In a clump of weeds, the three said, they encountered a creature like nothing they had seen before."

It was also not a gray of the type that has been autopsied. It had three "knobs" on its head and huge red eyes. But its skin was oily-looking, like the grays without their coverings, and it emitted a terrible stench which is described in a British documentary on the subject as ammonia. (Available on YouTube as "Varginha UFO Incident—Full Documentary.")

What unfolded in Varginha was apparently part of a larger event, and one for which the Brazilian military appeared to be prepared. Among Brazilian investigators, there is some suggestion that the North American Air Defense Command, NORAD, might have warned the military that the object was headed for the area where it fell. Nothing can be proven, but the fact remains that the military does seem to have been prepared to embark on a search mission of some kind in the Varginha area that night. They also delivered an injured person to a local hospital, but said that there was nothing unusual about the case.

I think, at this point, with so much official lying exposed, we can safely ignore the false statements of military organizations which are trying to cover up situations that appear to be dangerous and which they don't understand and can't control.

The event that began the case involved the sighting of a strange humanoid creature by three young women who were walking along a road on the outskirts of town. These three girls, Liliane Fatima Silva, her sister Valquira and their friend Katia Xavier, saw a small creature slumped against a wall. The red, slanted eyes and the three knobs on its head made it immediately obvious to the girls that it was no ordinary creature.

Dr. Leir conducted an interview with Valquira, who described seeing a short creature, crouching and motionless. She described it as "horrible looking, brown in color with an oily skin and a large head." It had what "looked like a set of veins bulging out of the skin on the shoulders and on the neck." She also described the three bumps on the head. She had only a glimpse of its eyes before it turned its head away and faced the wall.

We never think of more than 'the creature.' But the turning away of the head means that it either did not want to look at the girls or did not want them to see its face. And it was not simply an 'it.' There was a person there who, after all, must have been very far from home and well aware that its situation was desperate. As we shall see, this individual might have intentionally martyred themself for us. Even if it did not do that, but was simply the victim of a tragic accident or even hostile action, it nevertheless left us with a valuable gift.

It is so important that we see these others in as close to the richness of vision with which we see one another as possible. It isn't easy, of course, given that they are smelly, ugly creatures, and many of them aggressive in bizarre ways. I am not suggesting that we love them, though, but rather that we see them as best we can as the complex, nuanced beings that they must certainly be.

When we call them names like "biorobots" what we are really saying is that we don't think that it's necessary, or even possible, to understand them. We must not continue with this, because if we don't understand them and they are hostile, then they will prevail. If they are not hostile, it is even more important that we attempt to achieve a meaningful understanding of them.

Later on the night the girls ran from the creature, two military police officers on patrol observed a similar creature that appeared injured, and stopped to investigate. When one of them, Officer Marco Eli Cherese, just 23 years old, participated in a capture, the creature resisted mildly, causing Officer Cherese to handle it as he got it into the car.

Within two weeks, he was dead. An effort made by local investigators to find out more was, for the most part, unsuccessful. When Dr. Leir was in Brazil in 2002, he attempted to interview the officer's wife, but she would not give any details beyond the fact that her husband had indeed died in the hospital.

The British documentary mentioned previously includes an interview with Officer Cherese's sister, Marta Tavares, who reported that her brother came down with a fever, lost his appetite, became dehydrated, bled out and died. These symptoms are consistent with an extreme infection. The sister said "he was involved completely with the police operation from start to finish. He accompanied the creature from where it was captured to the Humanitas Hospital. He didn't get home until very late at night. When he actually captured the creature, he could have accidentally touched it." Tests revealed that Cherese's blood was significantly contaminated, containing 8 percent of unknown toxic substances. After the young man died, government officials demanded that the family cremate the body immediately, which was done.

It seems likely that the young man was contaminated by something that was associated with the creature, either a toxic substance or, what seems more likely, some kind of disease vector to which he had no immunity and which was communicated by touch.

If we ever end up with numbers of nonhuman beings or hybridized people with telepathic abilities moving about openly in our world, it will be important to be aware of this danger. Of course, the reverse is true, too. They might be as vulnerable to our pathogens as we are to theirs.

When the being was observed at the Humanitas Hospital, it was seen that it had a broken leg. Orthopedic surgeons were directed to set the bone, which they did. (I might add, here, that careful procedures were used to prevent contamination, and none of the doctors got sick.)

When Dr. Leir met them, he found them to be reticent in the extreme. They had received serious warnings from the military. For this reason, they are not identified individually in the interview recorded in Dr. Leir's book, but only as "MP," or medical personnel, and whether they were the individuals directly involved in the surgical repair is left open to question. However, Roger does remark that he noticed a change in the attitude of the individuals involved as he talked with them. They first claimed to be third-hand witnesses, but when they talked, it became clear that they were actually firsthand witnesses and participants.

The first question that comes to mind, of course, is how human surgeons could hope to accomplish a fracture reduction on an alien. It should be remembered that the biology document discussed in the previous chapter notes the endoskeleton is "similar in composition to ours," which would explain why it was possible for the surgeons to succeed. While the being at Varginha had a significantly different appearance from the ones that are discussed in the biology document, that the surgeons succeeded can only mean that the skeleton was similar enough to ours for this to occur.

The eyes of the Varginha alien "were large and red and staring at the ceiling with a blank stare." (This is like the 'dead eyes' look described by Corina Saebels and others. But when eye contact takes place, everything changes, and dramatically. It is unforgettably powerful.)

Were these eye coverings that were red instead of black? It would not appear so. "The eyes were large, slightly upturned toward the lateral aspects, oriental looking. They were red in color and looked like two glimmering pools of liquid."

The doctors did not want to look into the eyes. My experience of this is that you cannot help but look into them if that is wanted of you, so the blank stare they saw suggests that this creature knew what would happen if they did, and probably understood that the result would be that its leg wouldn't be set. It was as if the person behind the eyes was concealed, even though the eyes were open.

I can scarcely imagine what it would be like to end up in a surgery on a distant planet, surrounded by huge beings whose grinding physical voices would make it certain that there was no way to communicate with them.

Instead of looking into the eyes, the doctors continued about their business of examining the creature preparatory to their attempt to reduce the fracture.

The being is described as having four fingers without a thumb. In the biology document, four fingers are also described, but no reference is made to a thumb. As I quoted from *Communion* above, one of these fingers does work as a thumb, in my experience, but there is no way to tell which one just by looking at them. I wouldn't be surprised if they could all work in opposition, but my impression was that one of them was favored for this function. This is why I described it as I did in *Communion*.

The doctor described the fingers as "strange and different from human fingers. The creature was able to move each of his fingers so that they articulate with each other, and by doing so, could probably perform all the functions we could with the use of our thumbs. We were not able to tell whether these fingers were multi-jointed or, for some reason, the bones were flexible ..."

The biology document would seem to agree. The finger muscles being located "entirely in the hands" would give them this flexibility, and also account for the snake-like movement that the little boy saw in our house. The three sources, then, agree.

The size of the individual (just below five feet), the fingers, the power of the eyes, if not their appearance, all comport with both my observations and the biology document.

The descriptions of the blood are similar in both documents, but not in every respect. From the biology document: "The blood itself is also analogous to that of a human." From the Varginha interview: "When the blood was examined under the microscope, we found the cellular structure to be very similar to human blood, with the exception of the platelet count being much higher." From the biology document: "Platelets are present, but in smaller proportions than in humans ..." The interview states that the blood was red, but the biology document describes it as brown, probably because of the presence of so much metabolic copper. This is somewhat odd, by the way, as the document also states that the oxygen transport mechanism is hemoglobin, which is iron-based. Some animals, such as horseshoe crabs, have hemocyanin instead, which is copper-based. It seems possible that the transport mechanism is indeed hemocyanin, but that the author of the document was not aware of the difference.

As will shortly be seen, the difference in platelet counts could be important. Could platelets, so crucial to healing, be consciously generated by an injured being seeking to heal themself? That would seem, as we shall see, to be a possibility here. The doctor responding in the interview noted that the blood would "coagulate immediately" upon being exposed to air. He speculates that this might have been because of something about our air, but since it was not in respiratory distress, it is more likely that the high platelet count was a response to its injuries, and therefore that the ability to alter the count might either be a natural adaptation that is triggered by a wound, or something that the creature can do intentionally.

Since the Varginha interview was the only place that an anatomical description of an alien existed prior to the appearance of the biology document, could it have been used to fake the document? The important and also logically consistent discrepancies between the two documents would suggest that this is not the case.

While the description of the neck being quite narrow is present in both documents, the appearance of the head, as described in the Varginha interview, is radically different from the description in the biology document. What is not different, though, is the effect of the eyes on human observers.

I want to return, briefly, to them. One of the medical personnel interviewed by Dr. Leir specifically comments, "For some reason, all of us did not want to look into this creature's eyes and refrained from doing so." If contact evolves to another level, we are going to have to look into those eyes, and thus also deeply, deeply into what we hide within ourselves and, to some extent, from ourselves.

Now we come to the core of what happened in Varginha, potentially the beginning of a new direction for all of us.

After the surgery was completed and the bone was set, the doctors were still "highly tense" because they weren't sure if they had succeeded. They knew that not only were they operating on an alien, but for the first time in any of their memories, their operating theater was being guarded by armed soldiers who were watching their every move.

"Suddenly, out of nowhere, the room began to fill with a greenish mist." Of course, they were afraid that it was toxic. The military had locked the door and one of the nurses "began frantically banging" on it.

They saw that the mist was emanating from the creature. The doctor narrating this part of the interview "approached the head of the table. Without consciously realizing it, my gaze caught the eyes of the being. His eyes were glowing red and appeared as two swirling pools of liquid. They were pulling, pulling me in deeper and deeper."

"Information came pounding into my head. These were like thought grams, large blocks of information. Over and over, like someone hitting me in the head with a hammer." He became dizzy and nauseated, which is a typical reaction when this actually happens to somebody and isn't a fantasy. His headaches continued for two weeks.

The doctor was reticent to tell Dr. Leir everything that was in the "thought grams," but he was willing to say what the creature told him about human beings. The being said that its race felt sorry for us. "First of all, humans have the same potential and abilities to perform the very same things his race could do." The being said that in cases of injury or disease, it wouldn't be necessary even to use a hospital. They could "either individually or joined together" produce any healing that was necessary. The creature also said that they felt sorry for us because we did not seem to realize that we were "spiritual beings only living in a temporary shell and were totally disconnected from our spiritual self."

The patient was then removed from the hospital in what the doctors regarded as satisfactory condition and was never seen again. One would hope that something compassionate was done with this brave individual, but the cloak of secrecy prevents us from knowing.

The great gift that it gave us is the knowledge that we have lost critical powers. But why did we lose them? And how can we regain them ... or can we?

These questions will be explored more fully in part two.

CHAPTER FIVE

THE GENETICS OF DESPERATION

The sacrifice that was made at Varginha seems to be part of a pattern that characterizes the culture of at least some of the beings with this same general anatomy. In Diana Walsh Pasulka's *American Cosmic,* she characterizes a site in New Mexico (not the Roswell site) as a "donation site" because useful materials have been harvested from it for more than half a century. If the Roswell crash was also a donation, then it included sacrificial victims as well.

If so, why would they be willing to sacrifice themselves? One reason must be a completely different attitude about the value of the individual. The DNA evidence presented in the Reddit document is summed up by the statement, "They are artificial, ephemeral and disposable organisms created for a purpose that still partially eludes us." The bodies, then, may not be as central to their experience of reality as ours.

Or are these bodies the only kind that they possess? I can recall the sense of a "good army" that accompanied the much more dynamic person who managed my abduction and later became an intimate, if elusive, participant in my life and the life of my family. There are so many things about her that speak of powers beyond anything we know, including such a proficient ability to dominate us that, despite all that was unfolding in our lives at the cabin, and the persistent sense of danger, we kept returning.

I think of her often, of course, and wonder about our own personal relationship, its meaning, and why it even existed. What caused her to fall in love with me, for she certainly did, and then to commit the acts upon me that she did? And what caused me to so enjoy her presence in my life? For just as she loved me, I had powerful, complex feelings toward her. I still do. There was an aesthetic limitation, though—she was far from beautiful, and I had in my bed beside me a lovely, warm wife for whom I felt not just affection, but a towering love that to this day enriches me, even all these years after her passing.

But this other woman is still with me, too, and sometimes the memory is very poignant. Why was I drawn to an ugly, dominating and emotionally aggressive stranger, so much so that, just writing about her, I miss her electric touch and the skillful, dramatic intimacy that she visited upon me? My relationship with her was involuntary—I think. If she came back right now, though—walked in the door—I know exactly what I would do: I would open my arms to her. Then I think of what she did to me and the tragic boy who may have been the outcome of our intimacy, and I consider that I am more deeply her victim than I am aware of. I think of Stockholm syndrome, the tendency of a captive to identify with their captors. Was I so dominated by her I became entrapped in this way?

The more deeply we explore our relationship with these people, the more the contradictions pile up. The Varginha creatures seem to have been on a sacred journey for which they were willing to give their lives. The woman who was intimate with me was a very complex person indeed. But what about her soldiers? The Roswell cadavers had no sex organs, and therefore must have been fabricated. As we shall see, they weren't even designed well. But then how is it that they had the huge, absolutely extraordinary brains that we found? They were badly designed biological robots that were blessed with a genius so transcendent that it granted them powers over that natural world that to us appear as magic. And yet they could be thrown away without a second thought.

When we say the word "biorobot," we assume something simple, a kind of basic creature there only to do the bidding of its controllers. But that cannot be the case here, the reason being those larger brains. The idea, therefore, that they are simple robots must be approached with caution. They are not simple, and while they are fabricated, they may not be robots in the same sense that we mean the term. In fact, they must have more proficient minds than we do. They also might or might not be conscious, in the sense of possessing the mysterious sense of inner presence that we think of as self-awareness. If they were indeed sacrificed, it might be because a lack of self-awareness made them expendable. Or perhaps something I once observed and described in *Transformation* (p. 30) offers a more realistic explanation. After seeing one of them open a drawer, I wrote: "In that drawer were stacks of bodies like their own, all encased in what looked like cellophane. Their eyes were open, their mouths wide as if with surprise. I did not know what to make of it. The oddest thing was the way the drawer was opened with a prideful flourish. I was being shown something the two of them clearly thought was wonderful.

"I think that they must normally exist in some other state of being and that they use bodies to enter our reality as we use scuba gear to penetrate the depths of the sea."

As I suggested early on, we seem to be soul-blind in our world, and see the body as the whole being. If this is not true, then it seems to me probable that the cadavers at Fort Detrick had the same significance to their now-departed occupants as the bodies in the drawers would have had to theirs: they were not the final essence of the being at all, but simply a tool to be put to use when it was necessary to penetrate into the physical world. But, as always, in this extraordinarily complex situation, there are contradictions. The bodies are peripheral and badly designed. The brains are magnificent. This makes sense if the cadavers were something worn, and not intrinsic to the beings that once inhabited them, and needed only a minimal presence in the physical world but also required a brain large enough to accommodate their complex nonphysical presence. But if so, then why are the hybrids that they seem to have created between us and them afflicted with some of the same deficiencies as the cadavers? It is as if they had no better model of themselves to draw from. But why not? The woman I knew must have been more complex, so why not meld her DNA with ours instead of producing defective hybrids? —which, it seems to me, is what has been done.

Whatever the final answer, reading the letters in the archive at Rice University, it becomes evident that the perceptual and intellectual gap between us is vast. It could be that their only option is to give us samples and hope that we can derive meaning from them that makes sense to us on our terms. This would explain not only the materials but also the choice of these particular beings who, in their eyes, might be considered expendable.

No creature sees the world directly. Instead, everything perceives what its nervous system assembles in its brain, then delivers it to the mysterious detection system we call consciousness. This means that nothing is going to enable, say, a human being to perceive the world in the same way that a dog or a monkey does. We can only observe their behavior and interpret it in our own way. We cannot know how the animal regards himself or see the world through his eyes.

Given that our visitors are so complex that even their robots would appear to have brains more developed than ours, and, as we have seen, must have very different ways of perceiving reality, they might have been certain from the beginning that they could not really understand us any more than we can them. Thus, the donations of materials, craft and bodies may be an effort to bridge this gap by giving us artifacts that we can explore for ourselves and understand on our own terms. And as far as the bodies are concerned, big, complex brains or not, they felt that the lives of these at least semi-artificial entities could be sacrificed.

What the victims thought of this we cannot know, but we know that, as we create more and more complex artificial intelligence systems, we are probably going to need to program a fear of death into them if we are going to want them to feel that.

Ironically, something like this sort of programming once happened to us, and there exists a record of it. That record is in Genesis 3:22: "Behold, the man is become as one of us, to know good and evil." That is to say, we became self-aware. I am not suggesting that we were programmed, but rather asserting that this is a record of a time in human development when we came to know ourselves as beings separate from the rest of the living world. As we will see, this happened during the most terrible time humanity has ever known. We were driven by a prolonged and extremely violent natural catastrophe to recognize ourselves as individuals. In order to survive, we had to. But what if our visitors were never challenged in this way? Then they might still exist as a communal consciousness and have no concerns about giving up bodies if it serves some larger need of the species.

The Reddit document describes certain things about their DNA that are useful to anyone attempting to find their meaning.

The document begins with an expression of surprise that "their genetics are like ours, based on DNA." It continues, "We imagine that beings from an alternate biosphere would have genetics based on a completely different biochemical system."

Superficially, that would seem to be a logical speculation, but I think it needs to be looked at carefully. The document mentions that "some genes correspond directly, nucleotide by nucleotide, with known human genes or even some animal genes." As this implies that these creatures contain human elements, it would seem inevitable that the information-carrying system would be DNA. Otherwise, there would be no way to share.

For anyone who has experienced abduction, it must come as no surprise that human genes are present in these entities. During abductions, genetic material is removed from our bodies. That is probably, in fact, the primary reason for the abductions. There are stories of what people think of as hybrids, and there is no reason not to assume that there are variations, some of which are more and some less human.

For example, the boy who showed up at our cabin looked fairly human but had a shocking ability to interact with the mind that we do not share. It was a cloying, intrusive horror, to feel another presence inside one's intimate being, to be frank, and unlike human experience as we know it. It was a completely incredible invasion of my privacy, blatantly lascivious. When he was there, the most intimate acts of my life would leap unexpectedly to mind, or rather, be pulled up from my depths. They weren't awful—in fact, very much the opposite, quite beautiful, at least to me—but they were *mine*. It was a nakedness unlike anything we know among ourselves. I could not tolerate it, and drove him off.

He also suffered from what appeared to both me and my wife as some sort of autism. He was not in any way a normal human being. I don't think he could even bear to be among us.

I am describing him because he was the outcome of a mind that has skill at genetic manipulation, and, I think, must have created him and, as I now know from personal observation, many others as well.

If we are dealing with a species that has no moral restraints on genetic experimentation, or a compelling reason to ignore such restraints, they might be creating many such creatures. I must say that I see, in the obvious difficulties being experienced by that poor, lost boy, a limited competence. So why do it?

It could be hubris, but, looking at the overall situation, I think it is more likely to be desperation. They are trying everything to integrate with us, and, at least from my personal perspective, failing.

The document goes on to say, "For these genes, there doesn't seem to be any artificial refinement but a crude copying and pasting." This suggests that sophisticated techniques already in use by us were not available to whomever created these entities, or were not deemed to be needed. This reflects the strange combination of primitive and advanced that characterizes the whole contact situation. For example, their pell-mell invasion of our homes to gather sexual material that took place from the 1960s through at least the early 2000s suggests the sort of hurry that is associated with desperation.

Why would they be desperate?

There could be many reasons, but the most obvious one would be that they need something that we have, and need it urgently. Could it be that they have run out of novel DNA combinations and can no longer generate individuals that have not already existed?

Given that it would take billions of years for us to reach this same point, it seems unlikely. And yet, the cadavers we have studied display a very simple genome. So, the answer here is a 'perhaps.'

Given their warnings about the dangers of nuclear war and environmental collapse, it would seem that they want us to survive, so I think that consideration should also be given to the idea that the abductions might be part of a larger attempt to establish workable communications with us, probably part of an attempt to promote more of a communion between the species so that they can affect our survival decisions more directly. The idea would be to create "bridge" creatures, containing enough of both their and our DNA so that their brains would have a perceptual system that included enough of theirs and ours that they could communicate with both. The boy, for example, had human eyes and ears, and thus probably saw and heard as we do. Because he was also telepathic, he could communicate as they do. However, like them, he could not speak, but only gasp and grunt, which was one reason that the experiment was a failure. His sounds were not subtle and emotionally charged, though. They were the sounds of a human being who had been rendered voiceless by a birth defect.

I often think about what it must be like for him, to be a living, suffering failure created by another mind. Does he hate them? Or, as he would have no life without them, does he look upon his creators as his parents? Or me—how does he see me? And I wonder as well, is he now dead, and what might dying have meant to him? What has been his fate? I long to reach out to him and somehow give him some sort of comfort. I would like to have been a father to him.

If all such hybrids are similarly voiceless, their lives must be quite hard. Even if they are not unvoiced, I would think that possessing telepathy would be quite difficult to bear, especially if one could not turn it off. I have mentioned elsewhere that the boy smoked constantly, and I have since seen others like him who do the same. At one point, they tried to get me to vape. The implication was that this would make it possible for me to become more directly a part of their community. I would welcome more access, but not if self-harm is the price.

The boy displayed obvious symptoms of sensory overload, and as nicotine provides sensory gating, this may be one reason they use it so heavily. He was suffering from such gross sensory overload that he could not even bear to look at another person. While nicotine doesn't quell auditory hallucinations—or, in this case, the actual thoughts of other people—its tendency to reduce the intensity of stimuli might make the voices easier to bear.

There is another possibility, and it is disturbing. It is that these people may have been intentionally bred to be autistic, without regard to the effect this would have on their experience of life. The boy displayed extreme nonverbal sensitivity and, I think, heightened pattern recognition. He certainly experienced shared mental states, because he shared some with me. He could penetrate right into my mind. What would happen to me would be that thoughts, usually of a sexual nature and not always appealing, would come boiling up from my depths and I would be made to feel on stage, as if I was narrating my own private fantasies to a complete stranger and could not stop myself.

What I think was happening was that his heightened sensitivity was enclosing me and including me. I was being sucked up into his autism.

A number of writers, notably Donna Williams in *Nobody Nowhere* and Temple Grandin in *Thinking in Pictures,* have described the way their perceptions enable them almost to integrate with the feelings of others. In *Songs of the Gorilla Nation,* Dawn Prince-Hughes describes a sort of hyper-empathic state present in autistics that might as well be telepathic.

Dr. Diane Hennacy Powell's book *The ESP Enigma,* confirms this. In it, she shows that numerous unvoiced people on the autism spectrum demonstrate unmistakable telepathy with their families, friends and caregivers. Dr. Powell suffered for this scientific heresy, initially having her license revoked. She regained it, but only after proving that she was mentally sound.

The case made in the book has recently been further supported by "The Telepathy Tapes," a podcast produced by documentarian Ky Dickens, which explores in detail the lives, experiences and abilities of a number of unvoiced autistic people.

This gets me back to the boy and those like him. They live, as far as I am aware, in communities of their own and present a wary face to the world around them. Given the fact that Dr. Powell's work has shown conclusively that unvoiced and autistic people may develop this ability, could it be that the boy and those like him have been intentionally bred to be this way? He and his peers are probably as they are, at least in part, because of flawed genetic engineering, but it seems also that they could have been crippled in this way on purpose, in order to enable them to communicate with the grays, who, as we have seen, do not have the anatomy necessary to form words but who are fluently telepathic.

The fact that our visitors provided us with information by crashing disks containing living occupants, took us in abductions and created hybrids with no apparent concern about their personal welfare suggests the possibility that they do not possess the same moral compass that we do, or that they perceive the situation as so desperate that they feel that they must put any moral constraints aside. But what situation? Is it their need that they are responding to, or ours? Or both? We know that we are in jeopardy due to the population pressure that is stressing our planetary environment. But we don't know if they are also experiencing some sort of crisis.

If they are desperate, and I do think that this is a serious possibility, that would explain the headlong and precipitate efforts they are making here. We assume that they must be more intellectually capable than we are, but given their hybridization failures, which I will shortly discuss in more detail, I would question this.

If they are so capable, then why is it, as the Reddit document reports, that their genome consists of "16 circular chromosomes"? In our world, circular chromosomes are not associated with complex life forms. And yet they are also described in the document as eukaryotes. In view of the circular DNA, this is a surprising finding. (A eukaryote is a creature with a membrane-bound cell nucleus containing its DNA. Such a nucleus can be very complex and is essential to higher-functioning creatures such as animals and human beings.)

On Earth, all eukaryotes depend on linear chromosomes. This is because their complexity requires chromosomes that can interact with each other in order to properly replicate and segregate genetic material and maintain genomic stability. Exclusively circular chromosomes are found in simple creatures like bacteria, which do not have a need for complex interactions within the genome. Circular chromosomes are so called because they are sealed off from one another. They do not have openings that enable interaction, for the reason that the creatures involved are too simple to need this. Circular chromosomes are also present inside some eukaryotic cells, but they are supporting structures, not primary ones.

The document does not describe the circular chromosomes occurring in eukaryotic cells that depend on linear chromosomes, but says that the chromosomes of the creatures are entirely circular. If this is correct, it may be explained in a couple of ways. First, if the creatures are designed rather than natural biology, circular chromosomes are going to be easier to manipulate because their lower interactivity means that it is going to be easier to ensure that they will be stable, and there are going to be fewer and less complex decision points as the genome is created. Second, the absence of the larger intergenic regions inherent in linear chromosomes mean that the organism is going to be simpler to design, although also less flexible. But if these creatures are purpose-built, perhaps that isn't an important issue.

In addition to being simpler to design, there may be another benefit to using only circular chromosomes, which is that they are inherently more stable because of the lack of loose ends. Among other things, this would reduce the potential effect of breakages, which would likely add another advantage, because the organism would be somewhat more resistant to radiation damage. If these creatures spend time in outer space, this might be an important survival feature. That it might be a useful adaptation is also suggested by the fact that there are stories of people being injured in various ways, such as tissue degradation, which happened to John Burroughs after he stood in close proximity to a UFO during the Rendlesham Forest incident in 1980. He experienced damage to his heart, and it took intervention by Senator John McCain to force the Department of Defense to declassify his medical records so that he could receive treatment for his continuing problems. His remains one of the few cases in which medical records have been classified by the Department of Defense.

It can thus be reasonably concluded that the chromosomes are circular in part because the greater durability that results makes this a useful protective strategy. Additionally, because these are designed creatures, the use of simpler patterns and a limitation to just 16 evenly spaced chromosomes mean an easier design process, probably needed because of the limited proficiency of the designers. It could also mean that there was a lack of need to design more complex creatures because their use-purposes had already been decided. If all of their intended uses were known, there would be no reason to add complexity to the design which might trigger unintended consequences such as unexpected needs or flawed capabilities or genetic errors that might lead to disease, or even an unwanted level of independent thinking.

However, the fact that the cellular structure is eukaryotic while the chromosomes are circular means that these creatures are likely to have problems. Eukaryotic cells need linear chromosomes to function properly, the reason being that circular chromosomes will lack points of communication between them. DNA repair is going to be compromised, randomly offsetting the greater stability. The absence of the mechanisms needed in eukaryotic cells to ensure that the cell cycle runs correctly will mean more problems. It could cause mitosis to be unstable, if it works at all. This will result in missegregation and an aberration known as aneuploidy, which is an abnormal number of chromosomes.

I mention this specifically because aneuploidy can cause visible symptoms, and I think that I observed some of those symptoms in the boy. Specifically, these symptoms would have been present if he had an extra X chromosome.

Depending on where the breaks occur, aneuploidy can cause a range of different syndromes, with five of them being the most prevalent. Of these, one in particular, Klinefelter syndrome, seems the most likely candidate in his case. I have since met other males like him, and they seem to me to have been displaying the same symptoms. Klinefelter syndrome is caused by a failure of chromosomes to separate normally during the formation of reproductive cells, which leads to the additional X chromosome.

This results in two things: First, there is a high rate of autism among boys who suffer this tragedy; second, they generally display symptoms that include delayed or interrupted puberty. As to girls, I have briefly seen two, but I cannot evaluate their condition except to say that they were uneasy when near me.

In *Solving the Communion Enigma,* written long before I knew anything about these genetic issues, I described the boy as follows: "This was no boy, but it wasn't a man, either. He looked like somebody who had ceased to age before puberty, and was now not a man, but a sort of weathered child."

I think that it's clear that I was observing delayed puberty, or, more probably, a puberty that had failed altogether. In addition, the syndrome is most often seen in boys born to older women. I recall asking the woman on the cover of *Communion* if she was old and getting the reply, "I am old."

Given both his condition and this fact, I think that the boy did indeed suffer from Klinefelter syndrome, at least to some extent. Additionally, he was severely autistic and, to make matters worse, could not speak. My experience of him, as I have said, was that he was telepathic. I refer back to the speculations advanced above: He may have been that way *because* of the syndrome, which would suggest that this may not have been due to a poor choice of genetic engineering at all, but the result of an intentional choice.

If so, there are all kinds of implications. Are these people being bred so that they can transmit information back to their designers? Do they represent a replacement species? Or is my first supposition correct: They are poorly designed?

Unfortunately, given my current level of knowledge, I can only offer these questions. I cannot usefully speculate about motives. Perhaps if I had fuller access to the record I could offer more, but not at this time.

No matter why they are there, this combination of flaws and abilities must be causing these people great suffering, and is as clear an example as I can offer of what might be the most important assertion in this book, which is that we must not regard our visitors as gods or saviors. They are a complex, brilliant, subtle and yet strangely limited presence. We have much work to do before we can determine their motives accurately, and therefore evolve useful policies that will define how our societies and governments seek to engage with them.

We must never forget that the European colonizers wrecked more than one sublime culture not because they were morally superior or even more technologically proficient, but because the technology they did possess was designed to subjugate. We must not be disheartened, as the peoples they colonized were, by the drama of *power over*. We have extraordinary cultures on our planet guided by subtle and proficient minds, and when and if our visitors begin to express themselves into our lives, we must remember this and trust ourselves, and value ourselves. They are neither demons nor gods, but only people, although very different from us, with the same motive that we have: They want to get all for themselves out of life that they can.

In addition to my other thoughts, a possible reason that we are seeing hybrids at all is that beings who are themselves artificial and too inefficiently constructed to be really viable are attempting to improve themselves by combining their DNA with ours. While I could never assert that they have *ipso facto* been unsuccessful, what I have seen, and what the Reddit document supports, suggests as much.

Let's step back and take a look at what all this tells us about them and their morals and their culture.

The first thing that the whole picture presents is an extraordinary level of complexity. On the one hand, we have a creature apparently willing to sacrifice itself for our welfare; on the other, badly designed hybrids abandoned and left to suffer in our world. The boy, after all, was simply left in the woods behind our house, with no support offered to him or explanation to us. No introductions, no instructions. This certainly makes it look as if they didn't know what to do with him and simply dumped him. On the other hand, he was protective toward us, and I have seen that from others. A startling combination of provocative and protective. I have since seen that combination often. To protect me, they don't warn me. Instead, they attack me and scare me off. I hate to say it, but that's exactly what we do to a dog who is unwittingly putting itself in danger. And why? Because we cannot communicate with the dog in a more nuanced way.

It is important, always, to remember that the central challenge here is communication, and for this reason it is going to continue to be difficult for us to tell the difference between hostility and protectiveness.

If the overall picture I have painted here is accurate, then we can conclude the following:

1. There is compassion toward us, which was shown at Varginha and is of enormous value to us, as we shall see in later chapters.
2. The cadavers are biological engineering, but limited in their flexibility and proficiency. They would appear to be purpose-built receptacles rather than independent entities like us.
3. Hybrids with humanoid bodies who possess powers such as telepathy but lack the ability to talk have been created. These creatures, or many of them, seem to be significantly flawed.

Taking all of this together, it is to be hoped that these are the objectives: first, to raise our technological level; second, to demonstrate to us that we have vitally important powers that we have forgotten how to use; third, to create a bridge species that can communicate with both sides.

To sum up, what we are looking at here—always assuming that the intent is not hostile—is probably an effort to create a foundation for contact. It is not the way we would go about it. It is flawed and in some critical areas a failure. These flaws and failures can mean only one thing: They are not old hands at this. They are feeling their way and, so far, not all that successfully.

But how does this comport with evidence, such as that compiled by Jacques Vallée in *Passport to Magonia*, that this presence has been with us a very long time, and is probably responsible for previous beliefs such as the fairy-faith of Northern Europe and, into the present, the stories of the *djinn* in the Middle East?

The most likely answer lies in the calculation of astronomers, based on the prevalence of planets in our immediate area of the Milky Way Galaxy, that there are as many as 21.6 sextillion planets in the universe as a whole. If it proves true that movement over vast distances is not actually as much of a challenge as it now appears to us to be, then, inevitably, we will have been visited many times—for the most part in the past, it would seem, by species that did not take a lasting interest in us. In addition, if there are other universes separated from ours, as Dr. Roger Penrose has suggested, by nothing more than a thin membrane that can theoretically be penetrated, then it becomes even more likely that our present visitors are acting like newcomers because that's exactly what they are.

Even so, the question must be left open. We must never forget that we are dealing here with ultra-high strangeness—and, as a matter of fact, there could be few things stranger than encountering a creature that has a brain consisting of four hemispheres rather than two.

Let's explore just what that might mean.

CHAPTER SIX

AN UNKNOWN MIND IS WATCHING US

If the Reddit document is accurate, and, as we have seen, that appears to be the case when it comes to gross anatomy, our visitors' brain is "tetraspheric," which means that, unlike our brains, which have two parts, or hemispheres, it is composed of four parts. But before we conclude that they must be much smarter, it must be noted that the document also states that the brains are only 20 percent larger than ours. Each of the four sections is thus smaller than each of our two. They are described as having more complex gyrification, or folding, than our brains, meaning that they have a larger surface area. This is obviously designed to compensate for their smaller size, but as the document gives no indication of the degree of gyrification, there is no way for me to tell if the four sections are equal to or greater than our two in neuron count.

If the document is describing the brain correctly, then it is a very sophisticated organ housed in a much more crudely designed body.

Whether they are more intelligent than we are is an open question. On balance, I would find it surprising if they were. One reason is that smaller sections mean less activity per section and therefore more need for communication between sections. With only a 20 percent greater size, there would have to be a massive increase in gyrification to offset this, perhaps even so much more that functional disorders would be caused by too many neurons compressed into too little space. But the more important reason that I don't see them as being *more intelligent* so much as *different* is how they act and what they do.

Possible donations aside, if there are real crashes, then they reflect miscalculations or an ability on our part to destroy them. If they are donations, then they mean that our visitors do not understand how to communicate with us directly and must resort to physical demonstration instead. Therefore, no matter which case is true, we are looking at an intellectual insufficiency of some kind.

When it comes to genetics, not only do theirs show evidence of mediocre design; the alien-human hybrids, at least the ones I have known, have genetic diseases. Given all this, I don't think it's reasonable to conclude that our visitors are more intelligent than we are.

In the past, the idea that each hemisphere of the human brain had a specialized function led to the "left-brain, right-brain" concept, in which the left brain was organizational and the right brain creative. In his book *The Master and His Emissary*, neurologist Iain McGilchrist points out that the two hemispheres actually share the processing load, in the sense that the capabilities of one support those of the other. Thus, the creativity of the right hemisphere is supported by the analytical thinking of the left, and together they produce new ideas and new variants on old ones.

The two hemispheres work together through the corpus callosum, with each one participating in its own way in the overall tasks of thought and action. Together, they provide all the creative and executive functions available to us, and enable us to understand and live in the world in ways that are useful and meaningful. We can address the world with intellect and physical skills of various kinds, mediated by our emotions and, as many, including me, feel, our spirituality.

They have some abilities which differ from ours in an important respect: For these abilities to work, the laws of physical reality must affect them differently from the way they do us. Or is it that they somehow alter reality? We shall unpack that.

These are the abilities I am referring to:

Group 1—Movement
--Ultra-fast travel, with and without a vehicle.
--Unlimited propulsion.
--Gravity control.

Group 2—Personal Effects
--Invisibility.
--Levitation.
--Self-healing.
--Form changing,

Group 3—Mental Effects
--Telepathy.
--Mind control.
--Removal of souls from bodies.
--Creation of false memories in others.

Some of these seem that they would depend on technologies that address the physical world in unknown ways, such as those in Group 1, which enable extremely fast movement and infinite propulsion. The other effects listed in Groups 2 and 3 could either be the result of the extra brain functions that are seated in the two additional regions, or they could be expressions of technology, or a mixture of both. As an example, we use our vocal cords to propel words into the air around us, but also have technological means of amplifying them and transmitting them to distant places. Similarly, the group 2 and 3 effects might involve technological enhancements of brain-generated abilities. As will be seen, however, it may be that all of these powers are primarily internal to the beings themselves, and not technologies at all.

One case of an attempted removal of a person's self, or soul, from his body, as reported in *Them*, involved the throbbing presence of what must have been some sort of machine above the victim's house. Did this involve the use of some energy unknown to us to drag him out of his body—being attempted in the first place because souls are something this mind can perceive and may want to harvest? If so, perhaps that sensitivity is seated in one of the tetraspheres, and is a capability that our brains do not possess, or, if we do, have lost contact with.

How these abilities work is usually dismissed as a mystery explained by our visitors' advanced scientific knowledge. And yet, as their genetic engineering is demonstrably inferior to ours in some key respects, I don't see why we should assume that their science is unknowable to us, or even that much more advanced.

Whether we are looking at technologies or mental powers, or a mixture of the two, they must all have the same thing in common: For them to work, it isn't necessarily going to require machines—as we require, say, the use of a dihedral wing and an engine to fly, or a match to start a fire—no, for these powers to work the laws of physics must be affected on a deeper level. Specifically, depending on the effect to be achieved, the way one or more of the fundamental constants of nature work must be locally altered. For example, the laws of physics do not allow for faster-than-light travel; there is no known medium through which thought can be transmitted directly from one mind to another; small objects cannot be propelled at extreme speeds for long distances; repair of gross injuries without medical intervention is impossible. In fact, none of these effects can be accomplished within the laws of nature as we know them. And it's not simply a matter of manipulating those laws as we do with our technology; it is a matter of changing them fundamentally.

For example, for telepathy to work, something has to happen that cannot happen in reality as we now know it. Similarly, if travel over interstellar distances is involved, something must be done to change the laws of nature around the moving object—or else, even close to the speed of light, vast amounts of travel time will be involved.

The thirteen underlying constants that control the laws of nature are what must be controlled for these things to be done. If they can be changed by mental action, this could conceivably have to do with the two extra brain regions. That the brain and mind might be the resource needed to change them is something we will explore in detail later, but for the moment, suffice it to say that the constants probably have to be revised or most of the effects listed above cannot happen, no matter what means may be used.

Among the three groups, group one—ultra-rapid movement, gravity control and infinite propulsion—must require changing some of the constants. In this case, the speed of light has to be removed as a limitation, either by vastly increasing it, or, what is more likely, defeating the limitation by somehow removing the object being propelled from its constraints. To accomplish this, a number of the constants would need to be removed or changed: Beyond the speed-of-light restraint, the Planck Constant, which determines the size of an object at which quantum effects apply; the Gravitational Constant; the Cosmological Constant, which governs energy density and causes mass to increase with speed—all would have to be changed, at least in the immediate vicinity of the object in question.

Perhaps, if the object could generate its own gravity, a mind inside that gravity field could create a physics exclusive to the object that would have rules that allowed whatever was needed to happen—to literally create, in and around the object, a universe that was ruled by designer laws, one of which would be that infinite speed is possible.

The alteration of any of the other constants would always also require the alteration of one in particular, known as the fine structure constant. It is 1/137, and if it was any other value, the universe would be a very different place, either hostile to life or less friendly to its emergence. This is because it affects how atoms interact, and thus also chemical reactions. Because it is what it is, those reactions are what they are. If it were different, life might or might not even be possible. Certainly, it would not be like it is in the here and now.

It also governs the strength of electromagnetic interaction between charged elementary particles. Specifically, it governs the width of the fine splittings between the spectral lines in atomic structure, which ultimately underlie the shape of everything.

Were it a different number, all the other constants would have to be different in order to conform to it.

What is so mysterious about it is that it appears to be an arbitrary number, which led the great physicist Wolfgang Pauli into a correspondence with psychologist and mystic Carl Jung, about the mystery of its origin. It is also why my wife Anne would say that "God is a mathematical formula." She was referring to the equation that quantifies the fine structure constant.

To accomplish most of the "magical" things that we can observe our visitors doing, it would be foundationally necessary to be able to have local control over the fine structure constant. This would involve being able to control quantum electrodynamics, that is to say, to control the magnetic moment of the electron, which is the strength of its magnetic field, and its orientation. Changing this would profoundly change everything. The nature of reality would be altered. It might be that exquisite control of this is what enables our visitors to engage in seemingly supernatural acts like appearing and disappearing at will, walking through walls, changing their form, and even generating independent gravity fields that enable them to become, in effect, the god of the little universe inside those fields, which, as they have their own gravity, are necessarily not part of the greater universe. When we see a UFO darting across the sky, we are not just seeing something that possesses a mysterious power source, but also something that is *outside* of gravity. Because it has its own gravity, it is its own universe.

A fish can never affect, understand or even imagine what exists outside its water world. We are in the same situation, but I do not think that this is true of our visitors. I think that they can see reality from what amounts to an external perspective, and manipulate it from there. Surprisingly, I don't think that we are all that far from achieving this ourselves. The reason is that there is evidence that we possessed it in the distant past.

Right now, we don't even realize that we have lost something. That can change, though, as we shall see.

This fish—the human mind—is soon going to be able to see the stream from the bankside, thus gaining a much more true, accurate and powerful relationship with reality. We will be able to see what is inevitable about the future and what is not, and to perceive our own history, and thus our meaning, much more deeply than we can now. As we are now, we cannot access our true history, which is deeper than the marches and the kings, the poems and the pictures.

Once we can see reality from the outside, we will also be able to understand how to affect it in the same ways that our visitors can. The prey-predator relationship will change into one of symbiosis, and judging from the story "Visitors in the Trees" as analyzed in *Them*, this is wanted by at least some of them. In that narrative, an entire family experiences contact over a period of two days during which one of them—the wife and mother—is methodically led from one level of relationship to another, finally observing what had started out as two human beings enter the company of aliens, then turn into entities who were themselves dimensionless—existing on a shadow-line between physical and nonphysical being.

What this story is telling us is, if we survive the next few years, this will be where we go. This will be the future of mankind.

Time will tell, but I must add here that it is also necessary to revise our vision of that very factor. The assumption that time is some sort of natural force, like gravity or one of the other basic forces, has always been, to my mind, questionable. For all the talk of time travel, and even a personal experience that suggested that it's possible, I suspect that time doesn't exist, in the sense that it is an effect generated by a hidden physical process, a sort of "time factor" that is present everywhere and in everything.

In a 2024 paper titled "A magnetic clock for a harmonic oscillator," Alessandro Coppo, Alessandro Cuccoli and Paola Verrucchi describe an experiment that suggests that time indeed does not exist, but rather that it is a side effect of quantum entanglement (Coppo et al. 2024). Their essential thesis is that time is only perceived by entities that are entangled with something that measures time. They postulate that, at the beginning of the universe when no such entanglements as yet existed, an observer would have seen reality as static and what we perceive as duration, or change, as a fixed condition of objects. (How that would look is anybody's guess, but I suspect that we are going to find out soon enough.) In other words, before entanglement, we would have seen the universe from the outside. If this state could be induced now—if it was possible to become disentangled from everything that enforces duration—every clock or sign of temporal change, as it were—we would, ourselves, be able to see reality from outside of time. Then the human mind would join the mind of our visitors on the outside of the world.

To do what they do, they must be able to amend the constants, thus conferring on themselves what seem to us to be magical powers, but which actually reflect nothing more than an ability to manipulate reality from this external perspective.

They are fish who can leave the river and walk on land, and we can learn to do the same. As I said, I think we have done it in the past, an idea which I will develop in the second part of this book.

If so, then we do not need a tetraspheric brain to gain these powers. In fact, our brains, with their large and extremely complex hemispheres, and what must be a far less encumbered communications system, may actually be better equipped to acquire and use these same powers than theirs are. Just as their bodies are based on an inferior genetic structure, the tetraspheric brain may be a compromise brought about by the difficulty in engineering the much more complex genetic structure that large hemispheres like ours require.

The second part of this book will explore the possibility that we had the same abilities in the past that they do now, but either lost them or had them taken from us. What we may be looking at in the overall situation, when it comes to contact, is that we are finding ourselves confronted by a species that is fundamentally less potentiated than we are, and is at once trying to get what we have by extracting genetic material from us, and also using secrecy to make sure that we continue to believe we are powerless to resist.

Not only are we very far from powerless, in the end it seems likely that we will be able to use the same tools they do, and probably more effectively, and I would assume that they know this and are uneasy about it.

If this is correct, then they are not here because we are weak and easy to exploit, but rather because we are a sleeping giant. The contradictions and conflicts that we see among them probably reflect the fact that they are divided about whether or not we should be allowed to awaken.

I say that we should wake up, and now. And we can. We have the tools.

But what might it mean? Wake up in what way? How are we asleep? Above all, what bell needs to be rung to cause the giant to stir?

We will explore many different ways that this has been done in the past, even the relatively recent past, and also traditions that come from the deep past and may have entirely unexpected validity. For example, in Vedic and Buddhist tradition, there is the concept of *siddhis*, or accomplishments, which confer additional powers on properly prepared human beings. In one form or another, they are present in all the great religions. Catholics call them charisms. In Daoism, they are integrated into Quigong, in Islam they are called *karamat*, in Jainism *labdhi*, in Tibetan Buddhism *dngos grub*. While many different powers are described by the different religions, they all include healing, levitation, instantaneous movement and/or bilocation. All of these are powers that the visitors display, a finding that we shall explore in a later chapter..

It must be remembered that the Varginha alien said that we had *lost touch* with such powers. It is safe to say that his assertion, combined with the siddhi tradition, hint that our brains possess all of the powers in two regions that our visitors do in four, and if all the siddhis actually can be enacted, then even more. In our context, what we lost when the ability to heal that he talked about was lost could be the *shanti siddhi,* which is a profound state of inner peace that promotes healing. But how far can this be taken? Was he in such a state, and that was why he could heal his own body? How do these states work, then?

The Varginha visitor used healing on himself. But what did he actually do? What was that gas that filled the room? How do seemingly spiritual powers connect to physical reality—in other words, how can the mind alone alter the constants, and thus change reality to suit its needs?

We shall return to the question of these powers in part two of this book. Prior to that, we need to find out more about our visitors' inner lives, most particularly their spiritual lives, if they could be said to have such a thing. But before we do that there is another issue to deal with—or, rather, to face.

CHAPTER SEVEN

SHADOW OF THE TIGER

Are our visitors predators?

By focusing on the cadavers, while at the same time bearing in mind that they do not reflect the whole presence, we can explore that. Given this, it would appear that the physical abilities that are displayed in the cadavers are exactly what a predator who had human beings as his prey would need.

We know that they use those abilities for the purpose of capturing us. The fact that they generally also return us would seem to make them more like harvesters, but that still doesn't quite classify them as protective shepherds.

An abundance of witnesses, including this one, report that they used to take sexual material from us. This happened from the 1960s through the early 2000s and seems to have trailed off. Of course, it's also possible that they have devised a better way of abducting us without leaving residual memories.

As to communication, they have indicated to many close encounter witnesses that they want our survival, so maybe their harvesting of sexual material is part of an effort to ensure that our species is preserved if we go extinct on this planet—a sort of seed bank, as it were. Maybe, also, they are trying to create a species that can be a bridge between them and us. Or, as I have speculated elsewhere, a replacement species.

It must also be considered that they may not be able to communicate their objectives to us clearly, and given their inability to vocalize, I think that this is likely. If so, then they have no more of a way to reach us than we do an animal in the wild that we anesthetize, abduct, and then do to its body what we need to do to ensure its health and survival. Given that people when kidnapped are given warnings about environmental peril and the danger of nuclear war, one motive is to encourage us to protect ourselves from extinction. As sexual material is also taken, it would seem as if there is a backup plan in place.

These look like the actions of somebody who is trying to save us and going down more than one path toward that objective at the same time.

Why do they want us to survive, though? Is it because we are just so darned wonderful, or is there another reason, perhaps one more directly aligned with their own needs?

To succeed in dominating its prey, a predator needs at least one of two things: greater intelligence than the prey, or some sort of physical advantage. A pride of lions uses their greater intelligence to outsmart their faster but less intelligent herd of wildebeest. A snake uses its stealth and speed to compensate for the fact that any mammalian prey that might come its way is going to be more intelligent than it is. To the antelope that is the tiger's prey, its bright yellow and black stripes actually camouflage it, for the nature of the antelope's color vision causes those colors to blend in with the greenery through which the tiger moves. To the antelope, the tiger appears as a shadow.

Even though our visitors seem for the most part to return us to our homes, as I discussed in chapter 9 of *A New World*, there is no way to be certain that this will always happen, as we have no control over the situation.

Contact is not, however, limited to abductions and the sorts of attacks that happened in Brazil. As I discussed previously, when I returned to our cabin after refusing to go with the visitors in February of 1988, I was plunged into the moment when, as a baby, I took my first steps. This was a memory, but not at all what we normally experience. It was perfect, as if my adult consciousness had been sent back in time to my baby body. My mother's desk, the chair nearby, the sun shining in the window—it really was like being my babyhood self again.

If such memories exist, could they be of interest to somebody with the ability to see the soul and perhaps exploit it? If they could be extracted from somebody, would that be desirable? But do they exist? Neurologist Wilder Penfield conducted memory studies in the 1940s and 1950s using electrode stimulation of various brain regions that evoked detailed memories on the order of what I am describing. Since then, though, attempts to reproduce his results have been inconclusive.

The one that was drawn up in me was very beautiful. It included the absolutely magical feeling of moving on my feet for the first time—I felt as if I had turned into an angel, but all I was doing was walking.

If I could relive that memory, I would do it. If I could live somebody else's memory of an event like that in such detail, I would want to. But if doing that would mean taking it from them,

I would not do it. Or would I? If I had the ability to retain such memories but possessed none of my own—if I, in effect, was empty—would I perhaps be tempted to steal from them in order to fill myself?

Of course, given the information available, the question is unanswerable. On balance, though, I think it's something we should be aware of. If our visitors are a brilliant but experience-starved machine, we might be a treasure trove for them.

If so, then what happened to their life experience? One possibility is that they never had any—at least, nothing enriching enough to satisfy their needs and longings.

This possibility might not be as far-fetched as it seems. Dr. John von Neumann, a physicist known during his lifetime as 'the smartest man in the world,' is thought to have been on the scientific team that studied the Roswell materials back in the late 1940s and early 1950s. During this same period, he became interested in self-replicating machines. His book *Theory of Self-Reproducing Automata*, published posthumously in 1966, details his ideas about how such machines might work. In the early 1950s, he speculated that such a machine, if it could be designed to replicate the members of a species, could be sent through the galaxy in search of planets where they might be able to live, and when it found such planets, to create new members of the species and seed them there. But what if that isn't exactly right? What if they aren't looking for habitable planets so much as for a richness of life experience? That wouldn't have occurred to anybody then, but given what we have since learned about the visitors' activities here, it seems worth exploring now.

Von Neumann also speculated about the degree of deterioration in its store of information that such a machine might experience during its long, long journey. He assumed that it would travel at a percentage of the speed of light, but no more. His conclusion was that some deterioration would be inevitable, and that, the older the machine got, the less perfect its reproductions would be, but also that this would have been anticipated, and therefore that a self-repairing facility would be built in.

If Dr. von Neumann was right, it seems logical to me that such creatures would almost inevitably be a mixture of biology and robotics. The robotics would be there because their creators would be aware of the probability of programming deterioration, and would therefore want to minimize chances of degradation by using as many non-biological components as possible. To further minimize the effect of deterioration over time, the designers would have decided to keep the genetics of the biological component as simple as possible while still remaining true to their objective of re-creating viable copies of themselves.

As for their creations, they would have access to whatever knowledge their creators decided to grant them, probably whatever they might need to survive. It might be that significant parts of the parent species' store of knowledge would be left out, but what this might be there is no way to know.

Combine possible intentional editing with the effects of degradation over time, and the copy species might find itself lacking in fundamental attributes of one sort or another, and seek to self-repair. Perhaps one direction it would go in would be to extract memories from a target species, for it would inevitably be a sort of parasite—incomplete, unmoored in experience of its own, a wanderer at once among the stars and in the empty chambers of its own starving heart.

It sounds awfully science-fictional, but this is a very strange and complex universe, and I don't think it's wise, given the fact that our visitors appear to be purpose-built to prey on us, that we should dismiss such speculations until we know more about their motives for doing this.

The way their eyes work, the configuration of their muscles, their speed, their ability to control the mind and communicate telepathically all speak to their ability to dominate us, and therefore also to their purpose, which may be to control us so that they can extract from us what they need and want, whether it be sexual material or the memories that we carry so deep within us that we can't touch them ourselves. These memories, arising out of the amnesia of infancy, are, if my experience is any example, exquisitely beautiful and profoundly nurturing. Of course, somebody without memories might want them, and if there was a way to acquire them, they might do this. If taking experiences from us that we have stored in our brains represents the only chance they have to experience anything approaching the richness of real life, then I think we can expect them to try very hard to do that.

It seems, though, that there would be better ways than stealing our life experiences. What about engaging with human culture and having life experiences of their own? But perhaps the communication problem interferes. Perhaps it means that they can't. It may be that the theft of sexual material from us represents an attempt on their part to make themselves whole again by adding our much richer store of DNA to their own limited supply. I might add, here, that, just as there is no convincing evidence that they are natively more intelligent than we are, only more knowledgeable, their obsession with secrecy suggests that they find us intimidating and feel vulnerable in our presence. If this was not true, they wouldn't need to hide behind the wall of secrecy that they have created around themselves.

The forcible theft of sexual material is bad enough, but the theft of life experience—if this is happening—is even worse. If, at the end of our lives, we have nothing to show for our years of struggle, then why have we even lived?

I understand that this is all very speculative, and I must point out that it might not even be possible to remove experiences from a brain.

Before we can conclude that we have an apex predator that seeks our experiences, we must find out if it is possible to remove core life memories from a normal brain.

If we knew for certain that it was, we would devise all sorts of methods of preventing it. On the other hand, as long as we are unsuspecting, there is going to be very little resistance. Phenomena like "missing time," when people can be blanked out as if on anesthesia, means that we cannot know just how deeply our visitors are penetrated into our world and our lives.

A creature preying on a species as smart as humankind is going to need both stealth and intelligence, plus an array of skills that we, as their prey, are going to consider unusual, or even impossible. It is going to be in the interest of the predator to not only prevent us from knowing that it is here, but also, as we now suspect this, from knowing why. It will be careful to conceal its needs, and especially its vulnerabilities.

This gets me back to the issue of the cadavers. They are such imperfectly designed vessels that I have wondered if whatever animated them is even native to the physical world. I have touched on the idea that they might be something like diving suits for a creature that is not normally physical, but which sometimes needs physical utility. To a Western mind, steeped in materialism, this seems impossible. Our profoundly physical life experience makes it almost impossible for us to entertain an idea like this. But what if the physical world itself extends beyond parts of it that can be observed? What if there's more that we haven't yet detected, because we don't know that it's there? Obviously, there are many energies that we cannot perceive—more are, in fact, invisible to us than we can see in the field of visible light. But aren't energies different from material? If Einstein is correct, matter is energy, as the familiar popularization of his ideas, "Matter is energy slowed down," would suggest. While simplified, this is essentially a correct interpretation of his ideas. In our understanding, though, energy cannot have coherence like physical entities do, and cannot do things like retain persistent memories. To experience reality as we do, the anchor of a body is essential—except, what if it isn't? Could coherent energy, organized into persistent, self-aware forms, actually exist?

If the cadavers were originally designed to be entered and left at will by such beings, then they might not only artificial, but also disposable, which would be why they have been left behind in crashes. Big, complex brains or not, if they are designed to be used only when needed, then they are also expendable.

But that doesn't explain powers like mind control, telepathy, levitation, mastery of gravity and the sort of healing that the Varginha alien was capable of. Not even a greater level of intelligence can explain these, and I don't think it's useful at this point to continue to ignore them or pretend they don't exist at all, or are caused by the application of technologies we have no hope of ever understanding.

We can understand these powers. We can learn to use them. For example, we have always assumed that the extremes of propulsion and speed that we see are caused by advanced technology. We never stop to think that technology might be only part of the puzzle, or not part of it at all.

If these things are being done by mind—the fourth mind—how do we even approach that?

To me, the place to start seems clear enough: We must look into the emotional and spiritual lives of these biological machines.

I am quite sure that they have an emotional life. I have seen them disappointed in me and afraid of me; I have seen them work tirelessly to help me learn and do research and write my books. As well, I have been betrayed by them and loved by them, and done both to them. Our relationship is like a marriage—complex, passionate, and deep, but also demanding and, in its core, unsure. I do not understand who I have married, and I am sure that there are things about me that they don't understand. But here we are, at once in love one with the other and in fear one of the other and both of the unknown, and not comfortable with any of that, not on either side. And yet, we go on together.

There are many people in government and among the wealthy and the churchly who want to control the phenomenon in some way. Nobody can control it, not even our visitors. If they could, things would not be as they are, with all the confusion and chaos, and the deadening fog of secrecy that hangs over everything.

They would like us to believe that they are invincible. But of course. This is what the European colonizers did, and all too effectively. In 1910, for example, just four thousand British civil servants ruled the Indian Raj, which had a population of 315 million people.

Similarly, I suspect that there are far fewer of them here than they would like us to believe. It's not inconceivable that there are only a few of them on Earth. If so, then they may indeed be as desperate as they seem. The more vulnerable they feel, the more dangerous they are apt to be, and even if they are not many here, they have such extraordinary powers that even just a small number of them are likely to be a dominating presence if need be.

Even if they are a small presence, we are going to need to understand their level of belligerence very clearly, and respect it. If we fail—or refuse—to do as they wish, could they overwhelm us? Would they?

The best way to get an idea of the degree of danger here is to try to see the world as they see it, through the same moral lens. What might be the moral imperatives and constraints of a brilliant, desperate machine?

Let's see if we can find out.

CHAPTER EIGHT

SO VERY ALIVE, SO VERY DEAD

Now we come to what I believe must be the most secret of their secrets. If I am correct, then it is the fundamental reason that they have come here, and the thing they most want to conceal. (I wish that I could say that I was certain of this, but I cannot. It is my best effort to understand, no more.)

It has to do with time—with ascending beyond it and falling out of it—and I think there is reason to speculate that our visitors may have done both. Their great knowledge has removed them from the stream of time, but all-knowingness has not provided them the freedom and the safety that they expected. They have, instead, ended up in a life empty of meaning.

They know so much that they cannot be surprised, and that is their agony, and, I feel reasonably sure, also one of the main reasons to explain their coming here.

I can offer only a little evidence to support the idea that they have ended up in this trap, but it is telling. One such scrap of evidence is their ability to offer me an absolutely perfect recording of the moment I first walked. I don't see how you could find such a thing unless you could search across the whole time of a life in complete detail. Second is how it feels when they lock eyes with you. Once one of those small, slight, insignificant-looking creatures does this, you feel *known*. You see what it sees, which is all you have known through time, all you have been. Perhaps it is drawing on your genetic memory, if such a thing exists, or maybe it is what in Ecclesiastes is referred to as the long time of the soul, you as you truly have been, your shadow stretching back across all your life and even the ages, which, in some strange way and at such a moment, seem also to be part of you.

If we are to understand them—the danger they pose and the one that they face—we need to explore the possibility that they are trapped not in time but outside of it, because it might reveal motive. Only when we have established this are we going to know where we stand with them. To do so, we need to determine how they regard themselves as part of the world. What do they believe that they are?

The Reddit document discusses this. The EBOs, it states, "believe that the soul is not an extension of the individual, but rather a fundamental characteristic of nature that expresses itself as a field, not unlike gravity. In the presence of life, this field acquires complexity." As life-forms gain in self-awareness, the complexity grows, until the field "begins to express itself through these sentient beings." The life experiences of the sentient beings enrich the field, until it reaches what the document describes as a "critical mass" and an "apotheosis" takes place. The main goal of the EBOs, the document explains, is this apotheosis. Superficially, the word means to make a person into a god. One reading of this is that a person's apotheosis makes them a deity. Deeper, though, the shadowy presence of something very different hides in the word. *Apo* in Greek also means "away." *Theosis*" means "from God." "Away from God," therefore is another way of seeing what is meant here. Satan sought to replace God, and to leave time is to leave the world that God made—as it were, to replace God's knowledge with one's own. Or, put another way, to leave the stream of time is to fall from the grace of always seeing the world as new. Challenging God is the essence of what it means to be a demon. Although, I must say, that identifying them with meanings derived from a simpler age is not likely to be all that useful. Rather than identify them with old beliefs, it is perhaps more useful to see them as unfortunates who, because of their extraordinary minds and great knowledge, have lost the taste of the new.

The document goes on to mention that this process results in "negative entropy," but the author does not explain what he means, so let's begin by examining the scientific meaning of the phrase. From a scientific standpoint, the concepts of negative entropy that are most important here involve biology and information theory. Entropy is disorder. The less energetic, formed and organized something becomes over time, the more entropy is occurring. It is the breakdown of something, the rotting of the apple, an angel's fall, or perhaps the gradual information loss of a von Neumann Machine. On the great scale of things, slowly but surely, everything is disappearing into the dark. But negative entropy is the opposite of that. It is life and growth. Positive entropy is the slow unwinding of the world.

I can recall sitting on the deck at Geoffrey's oceanside restaurant in Malibu with Anne and Shirley MacLaine, listening to them discuss conflict between the dark and the light. As the sun set into the Pacific, Shirley would offer the opinion that the light was an intrusion into the peace of the dark, which wanted to hasten entropy—which it did, in human life, by spreading chaos of all kinds. Anne would agree, but assert that the light would prevail.

In terms of physics, Shirley was more correct: The universe has an end, at which time, if the "big crunch" hypothesis is right, it will collapse back into itself. If there is not enough gravity left for this, then all the stars will burn out, the black holes will evaporate, and absolute dark will once again prevail. But on another level, Anne was right, because something is being built here—which is why the phrase "negative entropy" was used. This dry phrase can be seen as another way to describe the journey of life itself.

In biology, negative entropy means the proliferation of life. "Negentropy machines," which is another way biologists describe living organisms, increase their order and complexity by drawing energy from the world around them. In information theory, negentropy refers to the increasing depth and complexity of information. The emergence of artificial intelligence is one example. Right now, it is housed in vast server farms that consume tremendous amounts of energy. Soon, it will move in a limited way into robots, then later into much more efficient and memory-dense biological systems—artificial brains, in other words. In fact, the cutting edge of negative entropy on Earth right now is the intensification of intelligence. When engineered minds enter designed brains that are housed in artificial bodies, we will have created something very like what our visitors have done. In fact, our own creations are destined to be very much like the little gray people with the big eyes who we now find so awesome. We will create entities that are so well-informed that they live outside of time, just like the grays. When we look into the eyes of our own creations, we will see ourselves reflected just as we now do in the eyes of the grays. And we will do this soon ... if we survive the global cataclysm in which we are now entangled.

With the advent of quantum computing, we will become able to design programs that interact at the same level of complexity as our own brains, and even more than our brains, and confer on them not only access to all knowledge in our possession, but the ability to make discoveries on their own that will enable them to find knowledge beyond ours, and to think beyond our thought. It is likely that we will soon be carrying devices in our pockets that are more intelligent than we are, and interacting with designed biological-robotic entities that know us better than we know ourselves.

Will they overwhelm us in some way too subtle for us to understand, and is that what has happened to the designers of the beings who are now exploring Earth and exploring us?

I wouldn't be surprised if our visitors' tetraspheric brains are the outcome of an effort to bioengineer themselves into hyper-intelligence, and that their success in doing this may have also been their tragedy.

At the same time, I would think that our own advances in the area of artificial intelligence are probably one of the reasons that we are getting closer to open contact, which is dependent upon at least a core of humanity becoming able to relate to them without being overwhelmed by their superior knowledge and skills.

I think that this is a significant part of what they hope we will do. As T. B. H. Kuiper and Mark Morris so presciently hypothesized in a 1977 article in *Science*, "Searching for Extraterrestrial Civilizations": "We suggest that knowledge, in a general sense that encompasses science and culture, is likely to be most highly prized by an advanced civilization. This could be formal, codified knowledge, or experiences whose value we have not yet appreciated. Furthermore, this resource is one that grows with time. We believe that there is a critical phase in this. Before a certain threshold is reached, complete contact with a superior civilization (in which their store of knowledge is made available to us) would abort further development through a 'culture shock' effect. If we were contacted before we reached this threshold, instead of enriching the galactic store of knowledge we would merely absorb it" (Kuiper and Morris 1977).

In part, this would explain our visitors' extreme reluctance to appear openly among us. As Kuiper and Morris suggest, they may be starved for the new. While that is probably correct, there may be more to it. As I have said, my impression is that we are fixed in the stream of time and they are not. The reason that this could be true isn't vague and esoteric. Rather, it has to do with information and predictability. The more information that is available to a mind, the more accurate its ability to predict the future becomes. If the information store becomes total, then there would cease to be a future. Everything would be predictable, and that would be why one of them showed me an image of a coffin when I asked him what the universe meant to him.

I think that Kuiper and Morris are wrong to suppose that they are here looking for fresh ideas from our fresh minds. I think that they are here looking for a very particular sort of experience: They want communion with us because they long to return to the stream of time. For us, every moment is new. If I am right about them, then they can no longer experience surprise, and they are here to share not our new thinking, but the moment-by-moment newness that is our experience of life.

For us, inside the stream of time, every moment is new. For them, every moment is inevitable, and that is no life worth living.

We perceive three dimensions and live inside a fourth, which is duration, or, as we experience it, time. If they understand entropy perfectly, and I think that there is every evidence that they do, then they live also outside of the fourth dimension of time.

As a result of the way we are, swimming in time like a fish in water, with no ability to perceive what lies ahead, we are like the inhabitants of Flatland in Edwin A. Abbott's 1884 masterpiece, *Flatland: A Romance of Many Dimensions*. In it, the two-dimensional Flatlanders can only see along a surface, with the result that they cannot conceive of three-dimensional objects. For them, a ball passing through their world starts out as a dot, becomes a widening circle, then turns into a dot again. They have no idea what is really there. When one of them discovers the third dimension, they reject the idea. Because it suggests that they don't actually understand the world, they become fearful. They try to stamp out their fear by executing him.

Our visitors are like someone who sees in three dimensions in a two-dimensional world: Just as the three-dimensional being sees solid objects that the two-dimensional beings cannot, our four-dimensional friends experience time as the fisherman does the river, as something flowing past them, not as a medium they live in and cannot see beyond. This means that they see present and past and future with the same clarity that we see only the present. If you ever have a chance to look into their eyes, you will feel the same anguish that I and others felt, as if they are looking into you more deeply than you yourself can. And you will be right, because they are seeing your whole truth, from the moment your consciousness began to whatever fate time and chance have in store for you.

Kuiper and Morris are right in this respect: They are here for the new, but not to discover new circuitry or materials or ideas. They are here to *experience* the new by sharing in our lives. If so, then the last thing that they would want would be to enable us to see the future like they can. They are not here to capture us or dominate us, but rather for the opposite reason: They want to share our freedom and our innocence. Letting them do that would truly be communion. It would also tame them to us, for they would never want then to lose us to fate or time or any catastrophe.

I wouldn't think that it's an accident that they are emerging into our world just at the moment that we are ourselves moving toward infinite knowledge. They want to enjoy our innocence while they can, and then, instead of deceiving us into continuing in this state, when we have mastered information, they want to participate in our experience of its loss. Then and only then will they be willing to disclose themselves to us in all their truth. I think that a final level of communion will then appear: The two minds, even though they both possess all knowledge, will see it in different ways, which will mean that both of us, precisely because we are, despite being infinitely informed, also different, will be able to explore one another's experience of timelessness with a sense of surprise that will never end.

I reflect back, once again, to those long-ago starlit nights at the edge of the world with Anne and Shirley. Anne was right: Life draws on energy to defeat entropy, and so gains complexity. But Shirley was, too: In the end, the darkness will prevail, but perhaps not entirely.

Separately, both of the sides in this strange dance called contact are destined to trip and fall. In fact, we're doing that right now. We are facing a danger of physical extinction, they the eternal coffin of their knowledge. But together, it is going to be possible to build something new—very, very new, in fact.

Let's explore what that might be.

CHAPTER NINE

THIEVES OF INNOCENCE

There's something that was not addressed in the conversations at Geoffrey's which our visitors have embraced and which the Western mind has rejected.

Our oldest religious document, which is the Pyramid Text carved on the walls inside the Pyramid of Unas, is a set of instructions for the pharaoh to use on his afterlife ascension to Orion, which involves the activation of an energy system along the spine. Since then, essentially every religious tradition has included an ascension method. Some of them, such as the one described in the *Tibetan Book of the Dead*, are intricate blueprints showing how to live and die, and how to negotiate the afterlife. The West has a much simpler method: be baptized into the accepted religion, then lead a moral life.

Starting in the Renaissance when the ancient Roman author Lucretius's great poem, *On the Nature of Things*, is discovered, and with it its secular vision of reality, Western thought begins to turn away from the notion that the real world includes invisible presences such as souls, ghosts, angels, demons, fairy folk, and all the rest. Driven by a desire to break out of the religious dictatorship that had ruled Europe, at that point, for nearly a thousand years, the Western intellectual community ends up entirely rejecting the idea that consciousness could exist in any way outside of the biology that supports it. In its effort to free itself from the doctrinal prison of the Church, Western thought has made the idea that biology creates consciousness as much a dogma as the Church made the idea that the creator is an immaterial being called God.

The document suggests that our visitors take quite a different approach, which is essentially the same as that of our old religions, but without any of the mystical trappings.

To understand the significance of this, one has only to consider the statement made by the being at Varginha: He felt sorry for us because we had *lost* skills such as the one he was using to heal himself. When you strip away the forms of gods and demons and angels and so on without entirely rejecting the reality they are meant to explain, what you have left are energetic states that appear to be quite real. Apollo or Ra or Yahweh or any of our deities represent powerful inner states, which we can see and use within ourselves if we cease to regard them as personalities external to our own being. Seeing gods as inner states that give rise to the folklore of deity will reveal their real power, which is hidden within us.

I would suggest that the old gods may be the rubble of a lost part of human experience, which could see life objectively, that is to say, the same way that our visitors see it, watching, as they do, from outside the flow of time. This is why we call the gods immortal: They are what the things of time—joy, sorrow, love, hate—look like when seen from the outside. They are, in other words, the Watchers of the Bible.

As we are, seeing the flow of life from the outside is as hard for us to do as it is for a Flatlander to see a ball for what it really is. Nevertheless, until we begin to understand gods as personifications of what were once powerful inner states, and what we now think of as religious ritual as the bumbling remains of lost practices of projecting the power of those states into the physical world, we cannot even begin to understand what we have lost and what our visitors have not.

The fundamental question of contact may be this: Can we share our experience of the new with them even as they share their total knowledge with us, without both sides losing their footing in reality? When they enter the time stream, will they forget the timelessness that is their home and their safety? When we leave it, will we lose the newness that is our reason for living?

In order to gain what both sides want from our relationship, we must somehow square the circle—but in opposite ways. If we manage it, that will be, in the truest sense, communion. We will have to admit the other into ourselves so completely, with such total surrender, that we will each become both.

We will taste of timeless knowledge but also continue to live in the new; they will taste of the new but continue to live outside of time.

To perform this ultimate act of union, what do we need to do and what will we have to give up? Are these misshapen, half-robotic creatures, deformed by the perfection of their minds, worth having as partners?

The reason we see them as demons and the reason we think of them as fallen is that they have, by escaping the stream of time, also destroyed their own reason to exist. In effect, they have stolen from God the right to know all. But we have done the same thing, to know ourselves. And therein lies the secret of a successful union between us: They must recover their place in the stream of time, and we must recover our lost powers that come from being partly outside of that stream. For both of us, it is a matter of seeking a new balance. By seeing the world through our vision, they can re-engage with the new. For us, opening our eyes to their vision will enable us to see and use the extra-temporal physics that they so adroitly command.

For both sides, there will be dangers and fears. We could lose our sense of the spontaneous and with it our reason for being alive. They could lose their vast knowledge, and with it their ability to navigate reality with anything close to their present power.

When we come together as a true community of mind, we will both achieve what we seek. We will leave time's river and look down, as they do, upon its golden days and storms. When they descend into it, they will leave their immensely informed communal being behind, and become children once again, naked in the mystery of the moment. In joining this communion together, we will both return to our childhoods. "Truly I tell you, unless you change and become like little children, you will never enter the kingdom of heaven," said Jesus in Matthew 18:3, and similarly in Mark 10:15, where it is stated that one cannot receive the Kingdom of God except as a 'little child.'

This is the deeper message of taking me back to the moment I first walked. For them to successfully navigate the time stream without sinking into it completely, and for us to successfully stake a claim along its shoreline without getting beached there, we must find our innocence and bring it, for there is no other way. The wide eyes of the visitor on the cover of this book reflect this. Beyond the shock of first seeing them, you will find in them the innocence that we both seek.

I wonder if it is really true that the blackness that we see is a covering? Even if it isn't and that is the eye of the unknown, we must still find in its reflection our own innocence, as they must find theirs in our time-enclosed gaze.

The document states that, after sentient beings appear in the evolution of a planet, the consciousness field begins to express itself through them, forming, as it explains, "what we call the soul." Sentient beings will, in turn, feed their experiences into the greater consciousness, increasing its richness and complexity. Eventually, there will come the "apotheosis" the document foretells, but then the author comments, "It's not clear what this means in practical terms, but this quest for apotheosis seems to be the EBOs' main motivation." Actually, it is clear what the term means in this context, so let's go a little deeper.

The idea, also central to the document, that the soul is not really an individual but part of a larger whole, reflects many of our own spiritual paths.

The Tao, for example, is conceived as an all-encompassing principle of being. In Vedantic philosophy, the Atman (individual soul) is viewed as part of universal consciousness rather than as a distinct entity. Jesus, of course, taught spiritual equality. This idea is also intrinsic to Gnosticism, which seeks, through study and discipline, to achieve gnosis—total knowledge, which is seen as mystical union with the divine. Buddhism is a journey toward understanding the self so completely that it is no longer the center of life. When asked about the nature of the self, Buddha very wisely remained silent. Our visitors did not: By knowing themselves too well, they have denied themselves the journey of discovery that is the key to enlightenment.

Despite the resistance of science, the idea of consciousness as a field also has modern proponents. Carl Jung has proposed the idea of the collective unconscious, which, if it exists, must be a kind of field in which we all share. Panpsychism looks on consciousness as fundamental to reality, much as the document suggests our visitors do, and there are quantum theories that view consciousness as a field, such as Giuseppe Vitiello's quantum integration theory, which proposes that consciousness emerges from quantum field dynamics in the brain. None of these are supported by invincible evidence, but one older theory is more proven. This is the von Neumann–Wigner interpretation, which identifies consciousness as what collapses the wave function. This means that reality is not determined until it is observed.

If this is correct, and so far the only time we have ever been able to measure a photon changing from a wave into a particle is when it is observed, then consciousness cannot be present only in physical bodies, but must rather exist in some more general way. If so, then the document is right to claim that "the field (e.g., consciousness) begins to express itself through these sentient beings," rather than the sentient beings originating it. When this happens, the soul forms in the sentient beings, which, I think, is much more central to the reason our visitors are here than anything in the physical world.

There are two great Western spiritual adepts, the fourteenth-century mystic Meister Eckhart and the twentieth-century Jesuit Pierre Teilhard de Chardin, who both had fundamentally important things to say about consciousness as a field and the nature of apotheosis that relate to the suggestions in the document.

First, let's explore how Teilhard's ideas relate—in fact, let's take a very careful look, because his idea, in particular, about consciousness as a field relates almost too well.

Here are some of the connections between Teilhard's ideas and the document:

1. His idea of noosphere, which he sees as a collective consciousness surrounding Earth.
2. His idea that, as matter becomes more complex, consciousness evolves.
3. His concept of the Omega Point, which would be the same as the document's "apotheosis," the ultimate evolution of consciousness to maximum complexity.
4. The foundational presence of love as the driving force behind the universe.

He also had many ideas that aren't reflected in the document, but I think that these are enough to serve as a reminder that the section about spirituality is the most questionable part of it. Superficially, it does not appear to be deep or particularly original. Look more carefully, though, and something changes.

The document can be seen not only in the context of many of our different religious and spiritual ideas, but pretty specifically in Teilhard's.

I cannot claim that the document's spiritual material is derived from human sources, but the precision of the connection to Teilhard's work means that this question shouldn't be closed.

The other great mystic whose work relates to the document is Meister Eckhart. Like Teilhard, he was a near-heretic. Teilhard was, in our gentler era, only silenced by the church, but in the brutal Middle Ages when Meister Eckhart was tried for heresy, if found guilty, he would have been burned to death.

Joy was the core of the Meister's teaching, and is the essence of apotheosis. One of his sermons contains what my wife regarded as the most revealing and deeply true statement in the Western religious canon. It is: "God laughs and plays." He said in another, "Do you want to know what goes on in the heart of the trinity? I will tell you. In its heart, the Father laughs and gives birth to the Son. The Son laughs back at the Father and gives birth to the Spirit. The whole Trinity laughs and gives birth to you."

This is very much in keeping with "have joy," not to mention my wife's belief that no awakening into higher consciousness is possible without laughter. She would say that, as soon as people became able to laugh at themselves, they were becoming sensitive to the presence of God. One of her favorite quotes hangs on the wall of this office: "If you want to make God laugh, tell her your plans." It was popularized by Ann Lamott.

And that's where we are on the journey to apotheosis: the planning stage. When and if this finally dissolves into laughter, we—and they—will be on our mutual way together. This may seem light but it's actually quite serious. We cannot go very deep into contact unless we can laugh—and that means both sides. Speaking of which, as a brief aside, the grays *do* have a sense of humor. What I can see of it is heavily sardonic, but it's there.

Speaking of time, because of the difficulty inherent in our different relationships to it, getting to the point of laughter is going to involve, on both sides, a surrender to the needs of the other that neither side as yet understands.

If you are outside of time, experience has no meaning for you, and believe me, getting close to somebody in that state is terrifying. You will perceive it as a danger worse than death. I know this quite certainly, as I have experienced that sense of annihilation. Also, my *Super Natural* co-author, Jeff Kripal, experienced it when sleeping in the same room with me at a conference at the Esalen Institute in Northern California.

I go to conferences there often, and sometimes the visitors will show up in the room I use, as it is relatively isolated and overlooks the Pacific, making for easy and secure physical contact for them. Because of this, people want to share the room with me, and one night when Jeff was sleeping in the second bed, there was a brief physical contact, first with me, then a few minutes later with him.

He experienced the sense of annihilation that comes from being drawn out of the river of time. He heard a great noise as if of glass breaking and then his own voice crying out from deep within himself, "Oh my God!" He was being drawn to the edge of time, and that voice within was his terrified soul, fearful of seeing so much of its future that nothing would feel new anymore, meaning that his life experience would provide little of what he came into this world to receive, which is the energy of discovery. The rest of his life would be like riding on rails.

We live in an illusion—a game, really. The future may be largely ordained, but as long as we don't know that, we gain self knowledge from our experiences exactly as if they were truly new. Having the illusion destroyed by being drawn out of the time stream is why we fear the visitors so deeply. Being cast onto the shores of the river is, for us swimmers, a loss of innocence so profound that it is literally a fate worse than death. Our visitors would seem to know this. This would be another reason that they are so careful not to come to close to us. In our coming relationship, balance is going to be absolutely essential.

Central to all of our cultures is something that must matter in a very different way to someone living in a timeless state: It is death. As they are going to know exactly when their bodies are going to die, they must surely see it more as we do the wearing out of a machine. Perhaps they simply grow or even buy a new body if they wish. Who knows how it might work?

Because they are on the shore and we in the water, we are going to have a great deal of trouble communicating. A Flatlander cannot imagine a three dimensional object like a ball. They have no information that would enable them to do so. But this works both ways—a three-dimensional being can easily understand lines and dots, but it takes a fierce effort of the imagination for him to conceive of a mind that can do only that.

Add together the fear that we feel when confronted with the threat of seeing too much of the future, and their fear of us lashing out at them when they dip into time and approach us, and you have the most challenging communications problem imaginable.

When they put on their bodies, they enter time, but I feel sure that they bring with them their knowledge of the future. To defend itself from the danger of such knowledge, the human mind will go into amnesia. Periods when we are outside of time cannot return with us to the time stream and so will fail to be recorded in the brain. This is one cause of 'missing time,' which, it would seem, can also be induced.

Our visitors slip into their bodies and strike out into the night, and draw us forth from our bedrooms just as the owl captures the little burrowing creature by listening to his underground rustling, ripping into his burrow and dragging him into the sky.

The chipmunk is devoured. Are we? Not physically, but I cannot let go of the idea that to someone in a timeless state, experience itself may be the ultimate aphrodisiac and the most incurable of addictions.

I will conclude this exploration of the many motives and possibilities raised by the spiritual comments in the Reddit document by looking at the relationship between its statements and our own traditions in yet another way.

A strange and yet persistent feature of the close encounter experience involves people being told that they are members of the family of the visitors who have abducted them. There is also the comment one of them made to a close encounter witness, Lorie Barnes, who worked for years as Anne's secretary. When they invaded her bedroom one night in 1954 and she recoiled in horror, she was asked, "Why do you fear us?" She replied, "Because you're so ugly." The response was, "One day, my dear, you will look just like us."

Admittedly, there is not much evidence here, but what there is does suggest that a level of relationship exists that we have either forgotten or have never understood. I have speculated elsewhere that we might be one stage of a multiform species. If so, this would explain their interest in us, and also their response to my complaint that they had no right to do what they were doing to me: "We do have a right."

Another possibility is that they got permission from some authority, the president or whomever, but I think it might go deeper than that. If the document does reflect their actual spiritual beliefs, then the match with some of ours suggests a hidden relationship between us that it is critically important to explore.

Over the course of this first part of the book, I have made numerous references to memory and the problems that it gives us when we attempt to determine which of our experiences are factual, which are mistakes and which are things that our minds have, in an effort to find some meaning, confabulated. There are also intentional lies, of course, but my experience is that most witnesses are trying their best to describe what happened to them as accurately as they can—and running into a great deal of trouble.

And then there is the matter of the past. Clearly, we have forgotten something very significant about who we really are and what our history has been.

In the next part of the book, I will deal first with the difficulties we have with memory, then explore the nature of what I think are the powers we have lost, and why, and how we can recover them by looking at our own minds, and our past, in a new way.

PART TWO

FINDING OUR LIBERTY

CHAPTER TEN

WE MUST REMEMBER THIS—BUT WHAT, AND HOW?

A great deal of our confusion has to do with memory. Human beings are not good at remembering things for which they have no prior reference. If we were dealing with contacts between two different social groups among us, or other earthly species, there could be confusion, but nothing approaching the disturbances to memory that are part of the close encounter experience.

In *Them*, I discussed memory problems, but here I would like to go a little further into the way memories are gathered and processed in the brain, so that we can determine the points at which they may be getting damaged by the friction involved in contact, and the points at which they may be getting intentionally altered or erased by our visitors—and how that might be accomplished. As to why they might want to intentionally erase some of our memories, I have discussed various possibilities previously in this book and others, but, in broad strokes, it must be because they don't want us to know what they're doing. I can't say for certain that this means that it's bad, but, when it comes to the removal of sexual material from our bodies, I feel that we do have a right to understand that, and should have the primary say in it, both on the collective level and as individuals. But until people know that this can happen, which depends on the lifting of the secrecy, I don't see how we can protect ourselves. Even then, it will be hard.

The greatest problem here, and, I think, the primary reason that our governments have reacted by presenting a hostile face to the visitors while also hoarding secrets, is exactly the same thing that affected Jeff at Esalen. When the visitors dip into time from the outside and contact us, it feels like a threat, maybe even worse than the threat of death. There is the suggestion of an even deeper annihilation. And this is not only a human response. I have seen animals in close proximity to the visitors, and their terror is extraordinary. This shouldn't be surprising. Any creature is going to experience annihilating terror when pulled out of the only world it knows into completely unknown conditions by creatures it has never seen before.

Also, from a human standpoint it is challenging to be face to face with somebody who appears completely different from anybody or anything you have ever seen, and who is possessed of strange powers.

This kind of fear distorts memory and can even erase it.

During the aftermath of the communion experience, I became interested in memory, and I have been concerned about the accuracy of my memories, and all memories, for that matter, ever since. And yet, since those early days, there have been a number of things that have encouraged me to believe that at least some of my memories are accurate. When I had my first experience with the visitors in December of 1985, there was little information available about alien abduction. I may have read about the

Betty and Barney Hill case in *Look Magazine* in 1966, but that was twenty years before my own experience took place, and I did not recall the details, at least not consciously. The appearance of the entities described in the article and what I remembered were similar enough, though, that I could have unconsciously connected them. I don't think so, though.

Still, I would suggest that the very nature of the memories of this sort of experience means that they must be kept in question—not dismissed, but also not accepted at face value until there is some corroborating evidence, as happened with the Reddit document and my memory of the visitors' means of evacuating waste.

The first thing to understand is that the brain may add imaginary material, which is what I assumed had happened to me when I saw that strange event. In that case, I was obviously wrong, but how do you tell?

Seminal research was done by Frederic Bartlett, which was published in 1932 in his book *Remembering: A Study in Experimental and Social Psychology*. He exposed British students to stories from other cultures, and found that what they remembered was changed to reflect their own expectations, based on their life experiences. That is to say, they used their imaginations to fill in parts of strange memories that they did not understand.

Subsequent studies have confirmed this: We don't remember past events as they happened, but rather as they coincide with information that we already have in our memory banks.

What happens, then, when we are exposed to something which has no relationship to anything we have known previously? The answer is that we distort it, bury it, or forget it entirely.

In *Them*, there is a story taken from *The Communion Letters* about a family who had a two-day group encounter. The work I have done on memory since the publication of *The Communion Letters* in 1997 has enabled me to show that this carefully orchestrated event was designed by the beings involved to enable one member of the family to remember every detail of what had happened. A process of incremental memory was used, where each event was related to the one that had happened previously, thus building a complete memory step by step. By the end of the two days, this person was able to remember something completely fantastic and totally detached from any previous experience in their life, except those they'd had during the contact itself.

What this tells me is that our visitors understand how human memory works, and therefore I think it's reasonable to assume that they must also know how to distort it and change it, perhaps even very fundamentally.

We need to be able to do this from our end, and the way to accomplish that is to build a database that we know to be real, physical memories of contact and train people in it.

It is possible to create just such a database. Let me be specific. Memories that are created due to physical experiences start in the senses. For example, seeing a bird flying past causes the deposit of a memory in the visual cortex, but imagining a bird flying past causes a much more complex type of memory, which will be stored in five different areas, including the hippocampus, the prefrontal cortex, the parietal and temporal lobes and the default mode network, where it will later be accessed during activities like daydreaming. The network becomes active when the individual is disengaged from external tasks, which are controlled by a different group of brain regions collectively called the task-positive network. It is drifting into the default mode state that allows the dreamer to detach from the ego, thus opening the door to more unfocused thought processes. The way the brain handles recollections of nonphysical events and those of physical events is different enough to make it possible to distinguish one from the other. It is also likely that detachment from ego has—or once had—much more profound effects than dreaming. As we shall see, a whole vast area of lost human abilities exists beyond the borders of ego.

There have been numerous studies suggesting that the difference between imaginary memories and memories of actual events is detectable, among them one in *Nature* in 2007, one in *Neuroimage* in 2009 and one in the *Journal of Neuroscience* in 2011.

What this means is that observing what brain areas become active when a memory is called up can enable researchers to determine whether it was imagined or resulted from actual sensory input.

Using this method, we can therefore build, by interviewing close encounter witnesses, a picture of what parts of their descriptions of what happened to them were recorded as experiences based on sensory input as opposed to imagination. All it takes is grant money and neurologists willing to do the work and to approach the matter honestly and objectively. Right now, though, there are numerous areas of resistance. First, granting foundations are generally unwilling. Second, the atmosphere in the scientific culture is not supportive of the idea that any such memories could be anything but imagination, meaning that objective researchers may be hard to find. Even good scientists are prone to researching to their assumptions. Third, the intelligence community and the defense industry may well seek to prevent any research that leads to the conclusion that the abductions, or any part of them, might be real.

At least, that is the situation as it now stands. Will it always be this way? Things appear to be changing somewhat, and the now widespread belief that an intelligent presence from some other place may be here and interacting with us is gaining some traction in the scientific and academic communities. And not a moment too soon, because right now what we have is an extensive 'alien' folklore that may or may not have a basis in fact, but which, if our visitors show up publicly before work like this has been done, is liable to become engraved in stone, whether it is correct or not.

One of the great mysteries of the close encounter phenomenon is missing time. I have experienced this myself on numerous occasions that I have noticed, and probably on others that I have not.

Let me cite two recent examples. I went to a wedding in another city that was also attended by one of the very few people still alive who had any knowledge of the bodies that were transported to Wright from Roswell. She and I had a brief discussion about them. The next morning, I flew back to Los Angeles. Because of my seat, I was the first person to get off the plane. As I walked down the jetway, I could hear the tramp of other feet behind me. I was sure that I would reach baggage claim first, or close to it.

When I got to baggage claim, it was dark. The conveyor was not moving. I wasn't surprised. I assumed that I was indeed the first. Time passed, but nobody else came. The baggage carousel didn't move. Finally, I headed for the baggage claim office on the theory that I had gone to the wrong baggage claim carousel and needed to be redirected. As I walked up, I noticed a distinctive blue roller bag that looked like mine. When I reached it, I saw that it was indeed my bag. Then I thought that perhaps it had been brought early because I'd been in first class.

I was absolutely flabbergasted when the baggage attendant told me that they had pulled my bag off the carousel after it had been there for forty-five minutes! When I looked at my watch, I saw that nearly an hour had passed between the time I stepped off the plane and the time I arrived at the carousel, a walk of no more than ten or at most fifteen minutes.

I then realized that I had experienced missing time. Not only that, it had happened in the middle of a busy airport.

It has taken me years of patient work, gently coaxing myself to remember, and I have to admit that I have not recalled much. I am fairly sure that a man began walking beside me shortly after I got off the plane. Then we sat at an empty gate. I don't remember what he looked like except that he was wearing a loose-fitting white shirt. He might have smelled of cigarette smoke. I recall that he was angry and demanding, and wanted me to quote every word that had been said about the bodies, which I was able to do. Oddly, I recall being eager to help him. Despite his attitude, it was not an unpleasant meeting.

I wish now to discuss an experience that took place in Paris in 1994. I have spoken about it a few times but never written about it. Anne and I had gone to meet our son and his fiancée. We had rented a flat in Montparnasse so that the four of us could stay together.

We arrived in the late afternoon, unpacked, then went across the street for dinner. During the night, I became nauseated, and when it was time to go sightseeing the next morning, I could not get out of bed. Just a few minutes after the others left, the doorbell rang. I was annoyed. I thought that they had come back for something they'd forgotten, and were forcing me to get out of bed rather than just use the key.

When I opened the door, I found two men standing there. They were very polite, and showed me what appeared to be official identification, saying that they were officers of the French intelligence service, and would like to come in for a brief conversation.

They didn't seem dangerous, and I was intrigued, so I let them in. One of them explained that they would like to induce missing time in me and interview me while I was in that state. I said that I would agree but only if they would leave me a recording. They replied that that was impossible, but that they would allow me to listen to the tape. The moment I agreed, the more talkative of the two said, "Would you like to listen?"

I agreed and—incredibly—he began playing the tape! About twenty minutes had passed and I had no memory of it whatsoever. None. It was just astonishing.

They warned me that I would not remember the conversation but for a few moments, which is exactly what happened. However, I do recall that it made me extremely happy.

By the time my family returned in the evening, I was feeling much better, and we went out for dinner again. Of course, I told them the whole story. We then had an early night because my son's fiancée had to fly at six in the morning.

When I walked down with her to help her get a cab, there the two men sat in a little Mercedes! I pointed them out to her. They smiled and drove away.

But there is a problem with this memory. It is that the only part of it that feels like normal recall is seeing the Mercedes. Were the men in it the same ones whom I saw the morning before, and did that event even happen? I think it did, and they looked like the same men, but the memory now feels somehow off, unlike the one of the missing time event at LAX, which seems completely normal. I did experience the missing time. But the memory of the man I met, and the conversation I had, is vague and indistinct.

I have cited these specific memories because they are exactly the sort that could be usefully analyzed with a functional MRI scanner, which would work not only with small, contained memories like those, but also with larger, more complex ones such as abduction experiences. If an efficient protocol could be devised and carried out with a significant sample of close encounter witnesses, the issue of what happens during these experiences could be resolved.

But why *does* it happen? While missing time may be at times something induced by the situation itself, it also seems to be at times an intervention. While it works like a short-term anesthetic, there is absolutely no sense of going under. The transition from one instant to another that can actually be minutes or hours later is completely seamless. The two men who visited me in Paris said that it was done with sound. I do not remember hearing this sound. The effect was just as I have described it above: I perceived no gap between the moment I agreed to undergo the procedure and when I was told that I could listen to my interview. I recall the gentleness of the man's voice as he questioned me, but, try as I might, I cannot remember a single word of what either of us said. As to the two men, I don't know who they really were or what, if any, government they represented.

The reason that I keep it in question is that I have been visited on four occasions by two people, always different, but always a team of two, one of whom spoke with me while the other stood back, I suppose to prevent any unwanted act on my part. All of them were in possession of advanced technology, which, if mankind possessed it, would inject significant changes into our technological and medical cultures.

The first team placed the implant into my left ear in 1989, pressing it home without breaking the skin. (In this case, it was the woman who spoke to me, and also in a very gentle, kind voice.) Then there was the French team, which consisted of two men, both very polite, who explained missing time to me.

The next one to come along, in Texas in 1996, was not so pleasant. This team of two used a device that generated a voice in my head and terrorized me with kidnapping threats. Not exactly gentle and kind—but in 2022, I once again saw one of the men who had done this. He was in a situation so very awful that, despite what he had done to me, I was moved to pray for him. Some of the people playing dirty tricks in this area would do well to look to their souls, I think.

Most recently, two men came to this apartment at four o'clock in the morning of September 22, 2019, two days before I was due to have my implant CT-scanned. They explained to me how it worked. What happens when it turns on is that I can see words racing past in a rectangular slit in my right eye. It was explained that these words, which move too fast for me to read them, are being drawn up from deep memory and made available in a more accessible level. This is why, when the implant is working, I have a much more extensive command of facts and details.

If sound is involved in inducing missing time, then infrasound (lower than the human ear can hear) might be what is used to produce it. This would be why I didn't hear it when the two men in Paris turned it on. Afterward, they showed me that they had played it on a small device that looked to me like an ordinary tape player.

Infrasound may possibly be used to disrupt memory, but I think the most likely mechanism that would lead to the missing time effect would involve the hippocampus. Studies have shown that infrasound can cause the hippocampus to become unable to capture events and consolidate them into new memories. A technology known as Transcranial Magnetic Stimulation can also suppress its ability to form memories, so this is another possible means of inducing missing time. It's not that the memory is distorted or suppressed, but rather that the memories do not form. Therefore, the person subjected to whatever causes this effect experiences a gap in time.

Any input that is stopped from being delivered to other brain areas for further processing, cataloging and retention is, quite simply, not remembered. But how would someone go about shutting down this brain area? Numerous anesthetics, led by propofol and ketamine, disrupt the activity of the hippocampus and interrupt memory formation. This is why patients have no perception of time during surgery.

Obviously, then, some sort of short-acting anesthetic could be responsible for missing time, but if so, how is it administered? There are many reports of missing time happening while people are driving cars and engaging in other types of complex activity. I have had this happen myself.

Could there be another mechanism?

There is a small amount of magnetite in the hippocampus, and overloading the organ with high-intensity magnetic fields has been observed to alter memory, but we don't have any studies showing that magnetic fields can have an effect as profound as shutting the region down.

The two men in France did specifically mention sound, so my thought is that infra, or very low frequency, sound is a possible mechanism. Infrasound can have numerous effects on the brain, and perhaps the right frequency and intensity might cause missing time.

While missing time is obviously a factor, traumatic amnesia also must play a significant role in the difficulty we have remembering close encounters.

In recent years, this type of amnesia has been confused with false memory syndrome, but the two are actually quite different. In addition, the idea that there is no such thing as memories that get buried due to trauma, once advocated by the False Memory Syndrome Foundation, has been shown to be incorrect. This idea was popularized by the press after a wave of apparent cases of hysteria in children were conflated into crimes by shocked law enforcement officials around the United States. People all over the country were arrested as Satanists after young children in their care accused them of engaging in abusive activities. An organization called the False Memory Syndrome Foundation worked to correct this, but, given that the founders were accused by their daughter, psychologist Jennifer Freyd, of abusing her, the advocacy of the foundation, despite support by renowned memory researcher Elizabeth Loftus, has always remained in question. What the foundation claims is that there is no such thing as repressed memory, because no mechanism for it exists. In her book *Betrayal Trauma*, Dr. Freyd disputes this. Because the original choice of a science advisor was so problematic and the accusations of the founders' child were what inspired them to create the foundation in the first place, findings associated with it are now routinely called into question. However, the foundation's assertion that hypnosis is, at best, an unstable method of retrieving suppressed memories has some merit.

While there is no real controversy over whether trauma can cause memory loss and lead to memory distortion—that is a well-established fact—whether or not real memories of traumatic events can be reliably retrieved using hypnosis remains an open question.

When Budd Hopkins first proposed to hypnotize me, I was not willing to have this done by a non-professional. To his credit and my everlasting gratitude, he drew Dr. Donald Klein into the picture. Dr. Klein was the Director of Research at the New York State Psychiatric Institute. He told me of his expertise in forensic hypnosis and said that it had enabled him to assist in the solving of seventy-two crimes as of our meeting in March of 1986. With such an impressive record, I felt sure that he would be able to help me recall details of what we both assumed was a criminal assault. When the alien abduction memories I had presented with only became more detailed under hypnosis, we were both quite surprised. Budd was, in my mind, vindicated for his belief, and I have always been grateful for the help he gave me.

At the time I was hypnotized, little was known about the abduction experience, and I had, as I have said, no conscious memory of reading about the Betty and Barney Hill case in *Look*. Had I not experienced two very obvious physical injuries during the December 1985 event, I probably would not have been willing to believe that anything had happened to me other than an exceptionally vivid nightmare. But the injuries were there, a needle mark in the side of my head and a torn rectum. The rectal injury gave me discomfort for more than twenty years.

Hypnosis has become a common method of getting detail from people who believe themselves to be close encounter witnesses, but I think that this deposit of narrative must be carefully interrogated by, for example, using the functional MRI scanning procedures I have discussed above, before we can determine what parts of it represent actual, physical experience and what parts are imagined.

I believe that human problems with memory go a lot deeper than our current period, I think that they go back very far in time, to a period known to geologists as the Younger Dryas, which commenced 12,000 years ago and lasted for 2,000 years. During that period, something so terrible happened to mankind that it imposed a kind of collective and generational amnesia on us. I think that this was when we lost touch with the powers that the Varginha alien demonstrated, and by so doing also lost touch with ourselves, our understanding of who and what we truly are, and, in the end, our true past, and with it our ability to become a cosmic species in the future.

But what did we lose? Like the brain, culture stores memory in many different ways and in many different places. To understand what we have lost deeply enough to regain it, I propose we take a journey down a hidden path, to a long-ago world that was very different from this—at once much smaller and more vulnerable, yet in many ways more powerful. To begin this journey, let's take a look at some of those powers as they have come down to us out of the mists of time, which may seem entirely unreal to us now, but which may once have been the only technology we possessed or needed.

CHAPTER ELEVEN

FRAGMENTS OF A LOST SCIENCE

There are many different types of myth and legend coming to us from our deep past, but two in particular stand out as being relevant to the possibility that something changed long ago which reduced the ability of the human mind to affect the physical world. One involves powers that we no longer possess, the other the devastating catastrophe that caused us to lose them.

I think that we may have literally been shocked out of using those powers, possibly because we came to fear that our use of them had actually caused the devastation we were experiencing. I very much doubt that our actions had anything to do with what happened to us, but the hardest thing for any intelligent creature to face is the fact that we live in a universe governed by chance. We would much rather have gods to control by our sacrifice and worship than the truth, which is that the greatest god of them all is randomness. And yet, this isn't quite as it seems, which gets me to a theory of reality that may explain a great deal about the difference between our experience of time and our visitors', which I have discussed above.

Block Universe Theory asserts that everything—past, present, and future –exists. Everything is permanent. In other words, the mysterious fourth dimension is not duration, but the entire huge block of reality, from beginning to end. Under this theory, time is just another dimension.

If this theory is correct, then what has happened is that our visitors have apparently escaped the thin line of moment—the now—that is what we perceive as we move through the block.

I think our visitors were probably motivated to escape from the block for the same reason that we appeal to gods—they wanted control over the randomness of reality. The ability to move and function outside of the block—outside of time—is the key to what appear to us as skills—levitation, shape-shifting, telepathy, spontaneous healing, instantaneous movement, and so forth—are actually side effects of their being able to address the universe itself from the outside looking in.

While these powers are rare among us and difficult to document with absolute certainty, so many people have observed our visitors using them that I think it's inappropriate at this point to reject them as impossible. These observations are not isolated anecdotes. They are integral to the whole body of testimony. The *Communion* letters archive alone, with its wealth of accounts of such powers being observed, is enough to make a good case that people are actually witnessing them.

"Nobody can read another's mind or engage in telepathy because there is no medium of transmission." No, there is no *known* medium of transmission. "Levitation is impossible because there is no way to control physical density." Again, no *known* way. "They couldn't get here because the distances are too great." Oh?

If you can leave reality itself, any and all of these powers become yours to control.

I think that we understood how to do this in the distant past, probably instinctually. Not only that, the method is probably memorialized in the Sphinx—not only the one on the Giza Plateau, but the many others that remain from earliest days, some of which also display wings.

This is one version of the riddle of the Sphinx: What has the strength of the bull, the courage of the lion, the mind of a man and can soar aloft like an eagle? It is a human being in a state of such profound objectivity about life that they are living outside of time itself—that is to say, they are in the same state as the visitors are in, with the difference that they are not trapped by their knowledge.

It is important to understand that this is not abstract. It is a real, physical and psychological state that we will recover when we have gained sufficient self-knowledge to achieve the balance that the Sphinx teaches.

So, two great signposts on the path we need to tread: recover the innocence of children and absorb the wisdom of the Sphinx. In this way, we can leave the river of time without becoming trapped, as our visitors seem to have done.

Proof that this state exists and can confer amazing powers has been hard to achieve. Abundant traditions worldwide attest to the existence of these abilities, with methods of attaining them found in practically every culture, and abundant descriptions of their use. But try replicating them in a laboratory setting, and they evaporate like so many mirages.

Six of them are all present in eight of the major religions. These are Hinduism, Buddhism, Jainism, Tibetan Buddhism, Islam,

Daoism and Catholicism. The powers are healing, clairvoyance, levitation, telepathy, astral projection (also known as out-of-body travel), and manifestation, or the ability to materialize objects.

Their existence has a long history in all these traditions, but they are prevalent in India, where Hinduism, Jainism and Buddhism all share similar beliefs. It is also India where the most fearsome stories of mythic battles in ancient times are found, but we'll get to that in a bit.

The earliest descriptions of siddhis come from some ancient Hindu texts, including the *Rig Veda,* which appears in tradition as long ago as 1500 BCE, but the *Yoga Sutras of Patanjali,* which appeared around 400 CE, is the first place where they are listed. However, a version of the power of ascension into higher realms appears in the previously mentioned Pyramid Text that is inscribed on the walls of the Pyramid of Unas, which has been dated to 2000 BCE. (I will refer here to the translations of this text that appear in *The Dawning Moon of the Mind*, by Susan Brind Morrow.) The text is a set of instructions intended to aid the pharaoh on his journey to the heavens. It has six sections. The first is an invocation to the stars, with special emphasis on Orion and Sirius, which are described as guides—not in the spiritual sense, but more as navigational guides for the soul as it rises into the stars. Aiming toward the Sword of Orion (now identified as Alnilam, the central star of the three stars just below the belt) is what opens the door to the sky. Next comes the falcon, which is a metaphor for the ascending soul. Then the instructions counsel how to navigate through the fire of dawn and into the dark where the stars can be seen. The soul, now able to navigate, aims toward the sword and so begins following a star map, eventually reaching Sirius and melding with it.

This is the earliest known description of the path of ascension, which later became the goal of virtually all religious practice, up to and through the early modern period in the West, and into the Renaissance. Modern Christianity is more life-centered, but the other great disciplines maintain this as their primary aim.

The Pyramid Text is a blueprint for reaching the stars, and one must wonder if it reflects something that was perhaps more than a spiritual journey. We will return to this when we discuss the true strangeness of the structures and artifacts ancient cultures have left behind.

Siddhis are acquired at the end of a long journey of enlightenment, and can be seen as the ancient path to Orion turned inward—which leads me back to the crucial question: why do the visitors possess them and we do not?

To gain some traction, let's look a little deeper, and to do that, we'll concentrate on the siddhis of India. There are thirteen, eight primary and five secondary.

The primary ones are:

Anima, the ability to become so tiny that one is invisible.
Garima, the ability to get extremely heavy.
Mahima, the ability to expand one's consciousness and being.
Laghima, the ability to become extremely light.
Prapati, or instantaneous travel.
Prakamya, the ability to get whatever one desires.
Ishitva, control over natural forces.
Vashitva, or mind control.

Among the secondary ones, which the visitors are also seen to possess, are levitation and telepathy. Others, such as clairvoyance, or the ability to see distant events; clairaudience, which is to hear distant sounds; and bilocation, they may or may not possess. I haven't witnessed them, nor do I know of any record of anybody else doing so.

Once you have seen a living creature change form or disappear or rise into the air, you are left with an intense and unforgettable question. But for those of us who have never witnessed such things, it's all too easy to just shrug and assume that the possibility isn't worth considering. That we might do such things ourselves seems doubly ridiculous: first, they're impossible; second, even if they aren't, we won't ever be able to learn such tricks.

I wonder about that, and I think that there is something hidden here that might provide us with some fundamental insight, and could lead in the direction of understanding these powers well enough for us to use them reliably, even to the point of being able to confirm them using the scientific method.

The message of Varginha is clear: We can recover what we have lost. Otherwise, why tell us at all?

Let's look at some of the siddhis. We might have a radar record of one of them, *prapati*, in use by the visitors. The word describes the structure of the particular type of movement that it records. "Pra" means moving forward, "pa" is getting something, and "ti" is a suffix that transforms the verb into the noun. *Prapati,* then, is an act of reaching forward, grasping something, then drawing oneself to it instantaneously. This is *exact*ly what the pharaoh was directed to do, by locating Orion and then drawing himself toward it. It is also the secret of how our visitors navigate, I would suspect.

This is a particular type of traverse, and it's important to be clear about how it differs from ordinary motion. It may involve technology, but I believe it's far more likely to be an act of mind, conceivably involving the visualization of the target location in such a way that a mind that can affect the material world with thought—can literally transfer instantaneously to the place which it is concentrating on. This would not be concentration as we understand it, a focusing of attention, but something deeper, the secret of which we have lost. But now that we know that we have lost it, perhaps it will be more possible to regain.

The following event may show the visitors using *prapati*:

During the 2004 encounter between pilots and personnel of the US aircraft carrier *Nimitz*, recorded both on the pilot's cameras and on the radar of one of the carrier's support ships, the *Princeton*, and which has been confirmed as anomalous by the US Navy, the object was seen to drop from an altitude of 80,000 feet to sea level in .078 of a second. This would be approximately 70,000 miles an hour. Although it would admittedly take over 9,000 years to travel a single light year at 70,000 miles an hour, given the distance covered, it was effectively instantaneous.

But did it move at all, or did we witness something more like what a Flatlander would see when a ball moved through the surface that he lived on? That is to say, were we witness to the part of a fourth-dimensional object that could be seen in our three-dimensional world? And is this what siddhis really are—aspects of fourth-dimensional events that are visible in the third dimension? When something from outside of time penetrates into its stream, the beings in that stream end up witnessing an event that defies the laws of physics as they understand them—thus instantaneous movement, defiance of gravity and unlimited propulsion. If this is correct, then these are not technological effects at all, but the results of the application of disciplines of thought that we do not now possess.

If our visitors are outside of time, they will be experiencing a very different physics from us. All the laws we know would still be active inside the envelope of duration, of course, but those outside it would also have access to laws hidden to us, such as the ability to pluck an object out of one moment and drop it into another, which could be what the Princeton radar recorded.

If the experience of time is due to our being quantum-entangled with objects that measure duration, as discussed in Coppo et al.'s paper mentioned above—that is to say, if time is an illusion caused by quantum entanglement—then it is probable that we can learn how to disentangle ourselves and rise above it, entering a state that is outside of spatial dimensions altogether. We can move our attention and awareness outside of time, just as the visitors do, and alter the position of objects from outside the rules of motion, just as they do. There are spiritual practices that move us in this direction, but what about current science? What about physics?

We have many theories of instantaneous movement, any one of which, if it proves workable, would deliver such mastery. Warp drives, electromagnetic drives, ionocraft, mach-effect thrusters and others have all been proposed. None of them are far advanced, at least in the public space, but one, described in a paper by Richard Banduric called "Breakthrough Spacecraft Propulsion Concept by Breaking Green's Reciprocity," might be a candidate for becoming a practical reality, and for a very unexpected reason. It relates to electromagnetic energy in a way that may parallel what a person entering one of the siddhi states may experience.

Banduric argues that charged elements, when in motion, will measure different forces despite interacting with the same electric field. This effect is called the breaking of reciprocity, and means that thrust could be generated from the imbalance. This thrust would not require expelling mass like a rocket, and therefore would enable unlimited acceleration.

Banduric presented his idea at the Alternative Propulsion Engineering Conference in December 2020. Similar to the wormhole concept, Banduric's proposal would enable the bending of space-time, but without the inherent instability of wormholes, as pointed out by Stephen Hawking, who argued that the negative energy needed to keep a wormhole open can never also be strong enough to stabilize it. The breaking of reciprocity does not result in a similar instability, but does require very substantial energy. Or does it?

We routinely observe UFOs that display both small size, sometimes just a few meters in diameter, and seemingly limitless propulsion. The one force we know of that might enable this level of functionality is the release of zero-point energy. Assuming that it exists and can be accessed, it would provide a limitless energy source. If we could power a system such as that proposed by Banduric, which defeats general relativity's light-speed limitation not with the deployment of a wormhole but by distorting space-time itself, with zero point energy, we would have something that could shift from place to place over limitless distances without, in the conventional sense, moving, but rather exiting reality at one point and re-entering it at another. But what does this have to do with some unknown power of mind? It would seem to require not thought, but technology.

It requires both. In his book, *The Day After Roswell,* Col. Philip Corso states that "Among the artifacts we retrieved were devices that looked something like headbands. These devices were a very sophisticated mechanism for translating the electrical impulses inside the creatures' brains into specific commands. ... The analysts at Wright Field believed that the sensors on the headbands corresponded with points on the multi-lobed alien brain that generated low-frequency waves."

Corso is too readily dismissed because some of his claims appear to be bizarre. Everything about contact is bizarre, and trying to cull the narrative for stories that are believable in our human context is probably a mistake. When I read, "the sensors on the headbands corresponded with points on the multi-lobed alien brain," what comes immediately to mind is the description of the four-lobed brains in the Reddit document. This means that Corso knew, over twenty years before that document appeared, that the brains were not hemispheric, and yet we live in a world in which all brains are hemispheric. I don't think this means that we should trust either the Reddit document or Corso's book completely, but it does mean that they should be taken more seriously than is now the case.

Eighty years of official lies and uneasy cultural compliance have led us into the trap of automatically assuming that even our own observations are not to be trusted, especially when something we are observing seems to fall outside the boundaries of the known. I am guilty of that myself, as I have mentioned previously. Seeing the beings in my front yard wiping the material off their arms is just one of many observations that I dismissed because they fall beyond the limits of the expected.

Don Schmidt and Kevin Randle report in *UFO Crash at Roswell* that "we finally got the former 509th Bomb Wing flight engineer Sgt. William Ennis to admit specific items about the craft ... He said, 'It didn't have a single moving part on it.'" This is also the impression I gained from my uncle, who I believe may have seen an intact craft.

If they don't have any moving parts, though, what enables them to fly? There is a book called *UFO Sky Pilots* by Grant Cameron and Desta Barnabe that contains accounts of close encounter witnesses who believe that they have flown UFOs. Their experience was that, upon touching a shelf or console in the craft, their consciousness flowed out of their bodies and into its skin—in other words, they themselves animated the craft. They could then make it fly anywhere they wished. Dreams, or memories of a technology that functions not only in the physical world that we know but also in some level of reality where physical and nonphysical meet?

Cameron says that he has interviewed nearly three dozen witnesses, who have all reported the same sort of experience. As Cameron puts it, "They became one with the ship and used their minds to move the craft around." He adds that "it is interesting to note that the stories are never exactly the same, but the fundamental story is the same...One person put his hand on a beehive in the middle of the room. Some put their hand on a ball on the console. Some put their hand over the ball but didn't touch it. Some put their hands on a flat console. Some put their fingers in holes in the arms of a chair."

Wonderful stories. *But how?* I can understand how it might be that the pilots are linked to the ship using telekinesis or some other process. We're working on brain–computer interfaces using helmets and implants right now. That type of communication between us and our machines will be ordinary in just a few years. But something more than moving matter with one's thoughts or brain waves is involved here. These people describe literally *becoming* the devices that they flew.

Recall from the Reddit Document: in the blood of the cadavers "metal ion levels are much higher (particularly copper) and glucose levels are significantly higher." Add to that bones filled with copper, and I think we might be looking at what propels the ships—the pilots themselves. As to the human pilots, it must be remembered that this experience is unfolding in ways that are very different from what we might understand or expect. Maybe their memories are of real events, and maybe they are something else. Right now, there is only one thing we should be doing with them, which is not dismiss them out of hand simply because they seem impossible.

Until James Clerk Maxwell theorized electromagnetic waves existed in 1865, nobody could imagine something like radio. And not until Heinrich Hertz showed they were real in 1887 did radio begin to be entertained as a physical possibility.

If the accounts reported by Cameron are at all accurate, perhaps it is possible to run the craft without understanding the underlying physics, in the same way that you can drive a car without knowing how its engine works.

This gets to a question, one that universally affects every traveler on this road: even if we master instantaneous movement, or something like it, how do we do these other things? How do we actually levitate, shape-shift, heal ourselves and conceivably also our planet and all the rest? How do we marshal these long-lost powers to save ourselves?

There are many traditions that speak of extraordinary energies. Among the most developed are those found in Daoism, Buddhism, Hinduism and Tibetan Buddhism. There are also references in Catholic mysticism to often extremely powerful ecstatic states, which will be discussed in the next chapter.

The field of neurotheology, which studies mystical states and religious ecstasy using the tools of neuroscience, has done studies of advanced meditators, some of whom have reported kundalini awakenings while being observed in a functional MRI scanner. What is observed is frontal cortex activity associated with increased awareness, and an increase in temporal lobe activity, which is probably associated with emotional intensity. Alpha and theta wave patterns change, with occasional spikes in gamma waves, which appear during episodes of acute awareness that practitioners might describe as being 'awake,' or more closely linked to a universal consciousness. Dopamine and serotonin changes have been noted in other studies, as well as "endorphin floods," which trigger euphoria.

When kundalini is activated, the practitioner feels energy moving up the spine. Some report that during this time, they perceive themselves to be floating, but there is no unquestionable photographic evidence of this.

There is a highly developed tradition involving a mysterious electricity in Tibetan Buddhism. The generation of Tummo, or inner heat, is an advanced practice that is used to burn away inner impurities. The abductee Betty Andreasson reported she experienced extreme heat during her primary close encounter, when a phoenix rose before her. The eleventh-century Tibetan Buddhist monk Milarepa is reported to have engaged in Tummo extensively, hoping to free himself from impurities.

Milarepa was also reported to have become able to levitate and to move great distances instantaneously through his practice of Tummo. It is now assumed that these stories are allegorical, but, given that our visitors can do these things, to me this is an open question.

Tummo works in the same basic way that most of the other energetic meditation practices do, basically by engaging with subtle energies that Western science no longer believes exist. Here, energy channels called *nadis* are visualized, then certain breathing techniques related to *pranayama* breathing (a yogic breathing discipline) are used to move energy through the *nadis*. This is accompanied by a repeated mantra and continued until a deep state of focus is reached. At this point, there is a careful examination of the state of the being, centered on re-visioning reality as a single presence rather than dualistically. Distractions like greed and anger are put to rest and past negative actions accepted, regretted and let go.

Tummo can become so intense that the practitioner's body temperature will rise. A study in the scientific journal *PLoS One*, which I follow, by Maria Kozhevnikov and colleagues showed that core body temperature could reach 101 °F during Tummo visualizations, which involve seeing bright heat within the body burning away impurities such as those specified above (Kozhevnikov et al. 2013). One method adds intense breathing to visualization, and it was the practitioners who used this method whose body temperatures increased.

This is not the only such study. While nothing as dramatic as levitation was recorded, profound physical effects occur during this and other such practices.

Most involve an engagement with chakras, but Western science, which is unable to detect subtle energies, has found no areas along the spine that confirm their existence. That doesn't mean they aren't there, though. As remarked above, they are first mentioned in the Pyramid Text where they are referred to as the seven "*ta-ntr*," so there must have been a connection between Egypt and India, a possibility to which we will return when we discuss the remarkable and mysterious Ed Leedskalnin, who created the Coral Castle in Florida, and seems to have been able to actually use what we now think of as imaginary magical powers.

From ancient Egypt through all the great religious traditions, that energy can be found buzzing merrily away, perhaps even before Egypt, used to enact the very sort of non-ordinary action that the Varginha alien demonstrated. It cannot have been in use in Egypt, because we have their medical texts, which are very far from magical.

We need to find our way back to that energy, but this time through the acute lens of our modern minds, not simply as something we use without understanding, but as something that we understand and can integrate into our knowledge base, and use consistently to live our lives in an entirely new way.

I would venture to suggest that we *can* make use of the same energies that our visitors demonstrate to us every time they race past overhead, or come as shadows in the night into our sleeping lives and float us off into their mystery.

Given the increasing instability of the climate, we could be in the process of entering a second Younger Dryas, in which profound and lasting upheavals may cause Earth to become far less able to support us, and not just for a few years, but for millennia. Extinction events are an ordinary part of the planet's geologic history, and given the changes taking place now, we could end up in a situation in the next few years where it has become obvious to everyone alive that spaceship Earth has come to the last phase of a titanic struggle, and is going to enter a state where it cannot support us in such numbers as are alive now, if at all. If this happens and our visitors will not or cannot help us, then we must face the fact that we are alone and are in danger of becoming lost forever in the dark and the stars.

What, then, shall we do?

CHAPTER TWELVE

AWAKENING THE SLEEPING BEAUTY

Can we master the energy that is, perhaps, behind all the siddhis? They exist, because the visitors demonstrate them. Therefore, a physics unknown to us that governs them also exists. We see its principles being applied every time a UFO zips past or someone is floated up into one.

If we could find just one siddhi and truly understand it, that might open the door to them all.

But which one? They all seem as impossible as flying at thousands of miles an hour with no propulsion system or levitating—or wait ... perhaps there is some evidence for that one.

Could something as seemingly impossible as levitation actually be our way into accessing this energy more broadly?

In the fairy tale "Sleeping Beauty," an evil fairy causes a princess to fall into a trance for a hundred years. She may be awakened only by a kiss from a royal prince. As the years pass, her castle is surrounded by brambles, until it has almost disappeared from view. But it is noticed by a group of hunters, one of whom is a royal prince. He forces his way through the thorns and finds the sleeping princess. Moved by her beauty, he kisses her and wakes her up.

I think we were enchanted long ago—by what, exactly, we shall discuss beginning in the next chapter. It certainly involved a great natural catastrophe, and perhaps something else, something much stranger. It diminished the power of the mind so completely that our memories of our greater past turned into longing for our lost powers, which we sought to restore through what started out as ritual, then became religion. Starting around 7,000 years ago, we began to worship elaborately personified gods, which we created to control an environment that has almost destroyed us.

This happened when decimated populations were struggling to recover from the catastrophe of the Younger Dryas, which ended about 2,000 years before organized states began to appear in India, Egypt and the Middle East. I suspect that the elaborate ritual processes that appeared along with those states, as well as the implements involved, may have been copied from tools that worked in ways that we had forgotten during the chaos of the two millennia that had come before. In short, the old religions, with their elaborate rituals, implements, ritual spaces and supernatural pantheons, were actually the shambolic remains of what was once a coherent system of influencing nature, a high science that did not work as modern science does. It required no physical infrastructure, no engines, no reactors, no power sources except—well—us, or rather, what we once were.

If this is correct, then the ancient religions were a cargo cult that sought to restore lost powers that were no longer understood.

To recover that understanding, perhaps it would make some sense to go back to the beginning of history, when, I believe, what

I have called *a science of the soul* existed. What I mean by that is that we once understood ourselves well enough, perhaps instinctually, to activate powers like telepathy and levitation at will and be able to evoke them whenever needed.

A good place to begin is to explore just one lost power. I have chosen levitation because it seems impossible, and yet the evidence that it can happen is compelling.

I was interviewing a longtime paranormal researcher Paul Eno about his book *Dancing Past the Graveyard* on Dreamland on August 26, 2022, when he recounted a story from his early life as a novice Catholic priest. He was assigned to be the assistant to a priest who was both chaplain at a local mental institution and an exorcist. He attended the exorcism of a seventeen-year-old girl who was a patient in the hospital. She had schizophrenia and was in drug rehab. Her Catholic family had requested an exorcism. During the ritual, as witnessed by the exorcist, a doctor, a nurse, an attendant and himself, the patient "rose out of the wheelchair not using her legs—she was still in the seated position—and Leonard and I (the other attendant) had to push her back down." In other words, she levitated. I asked him what touching her had felt like, to which he replied, "It was very electric."

These four words are a signal from our lost past, but they bear within them the promise of its rediscovery. This is because they reflect the existence of an energy that can be felt and is therefore a part of nature, and is connected to levitation. As there is an energy, there must be a physics that governs it. It is time to find a way to detect this energy so that we can discover and understand that physics—in short, come to understand *what it is*. I have theorized before that the energy might, itself, be conscious, and might therefore need to be related to very differently than we do other natural forces. If it is self-aware, we will not master it in the same way we do, say, light or magnetism, but rather will need to interact with it as we would with another—if distinctly elusive—person.

Looked at in this way, we can see that a vast number of rituals have been deployed to attract its attention and gain its cooperation—and at the same time, it has been anthropomorphized into gods, angels, demons, *djinn* and so forth.

However, when we see the strange bodies that have fallen into our hands, I think that what we are seeing is what it has itself devised in order to function in the physical world.

With the more orderly understanding of which we are now capable, it seems possible that we can evolve a new relationship with this energy that is perhaps more objective and doesn't see it through the medium of belief at all—no ghosts, gods, angels, demons or other magical beings, but rather, simply, *somebody else* who has left us a calling card in the form of the bodies.

If I am correct about this, then the establishment of a coherent and objective relationship with our visitors is going to turn out to be even more extraordinary than we have imagined. It will be a relationship with a whole level of consciousness that we have previously seen only "through a glass darkly," which is to say, through a filter of mysticism and folk tales.

Before Paul and a number of other witnesses, a young woman, largely composed of water encased in the intricate tangle of bones, nerves and flesh called a body, rose into the air. Therefore, she defied known physics. Whatever happened to her, it caused what is basically a big bag of water weighted down with masses of calcium to cease to be heavy, either that or to be lifted by an invisible force of some kind. Whether something in her physical condition changed, or she was lifted by a conscious energy that possessed that capability, her body floated out of the chair and had to be pushed back down, which was easily done by two people who also felt what they perceived as electric energy entering their hands from the levitated body.

What this means is that an object with a density of about 1.0 gram per cubic centimeter rose into air, which has a density of only 0.001225 gram per cubic centimeter.

Something that cannot happen, did. Therefore, there was a natural force, conscious or simply not understood by us, acting on her. Or is there some other explanation, one that keeps us properly on the ground?

There is certainly one: it's all nonsense, the reason being that it has to be. I would agree that we know of no force that can do something so fantastically impossible as to alter the density of a body to where it is lighter than air, or raise it by some invisible means. And yet—well, it appears to have happened rather frequently in the recent past, and a clear-eyed look at what has been reported can perhaps get us beyond not bothering to study that mysterious electricity because we assume it doesn't exist.

If it does, then it must affect the body and its clothing in some way, either entering it from the outside or being generated within it. In either case, it is going to be detectable, and perhaps the first place to look for it, I would think, would be in extreme states such as ecstatic trances. And indeed, there has been some neurological research into these states, one example of which is a study by Emma Huels of the University of Michigan and colleagues entitled "Neural Correlates of the Shamanic State of Consciousness" that appeared in *Frontiers in Human Neuroscience* (Huels et al. 2021).

The paper reports a change in the activity of the default mode network. When we disengage with our sense of self, or ego, during deep meditation or states like mediumistic or shamanic trance, it shuts down, leaving room for the imagination to flow freely, and perhaps also for something like conscious energy to engage with us.

With chant and dance and perhaps the use of drugs, a shamanic practitioner's ego can get so profoundly separated from the intrinsic self that it can feel as if another, entirely different personality has entered the body, as happens when a Vodoun practitioner becomes possessed by *Loa*. But is that what is actually there? Because the connection has been filtered through the beliefs of the entranced person, we cannot know, and this is what has to change. In order to invoke this state reliably enough to study, belief has to be replaced with something that might be called objective acceptance—open, neither questioning nor believing.

As the intensity of the experience increases, theta and alpha brain wave activity increases. The brain detaches completely from its normal state of self-awareness. It is at this point that other realities of many different kinds are perceived. Also, these realities can sometimes actively intervene, as happens when a medium goes into trance, or when, as with the patient Paul Eno observed, a woman becomes so filled with an unknown energy that her body itself and the clothing around it changes its weight enough to levitate.

It is in such moments that what is believed to be contact with the spirit realm occurs. But we only call it that. We have never understood it objectively, and only know the characteristics of the entities, which are, after all, only filters in our minds through which this unknown power manifests itself. The god of the saints, the *Loa* of the Vodoun practitioner and the electricity of Paul Eno's patient are, in all probability, either related or actually the same thing.

Surely, there is a way to detect and measure this energy consistently. But we never look at it as something that is looking back at us. We cannot succeed by limiting our detection efforts to the deployment of instruments. We need to engage with it, to learn ways of communicating with it that are reliable enough to enable us to ask for its participation in our experiments.

So also, phenomena like out-of-body travel, if they really happen, must be measurable. And, in fact, there have been several studies in which feelings of being separated from the body have been induced, some using electrostimulation of the brain and others virtual reality glasses. Even before these experiments, Dr. Charles Tart studied a woman (Miss Z) who reported frequent out-of-body experiences. He published his results in 1968 in the *Journal of the American Society for Psychical Research.* The study reported on an experiment in which Dr. Tart placed a random five-digit number in a location that Miss Z. could not physically see. He found that, after going out of her body and moving in the nonphysical state to a place where she could see the paper, she was able to accurately repeat the number (Tart 1968).

With five digits, the statistical probability of this being a lucky guess was vanishingly small. But, as seems to so very often be true when dealing with these cases, there was no consistent repeatability. More than anything, this capriciousness suggests to me that we may be trying to connect with something that possesses consciousness. In fact, given that our visitors may be capable of entering and leaving their bodies more or less at will, could it be that we are dealing with them when we attempt to engage the interest of conscious energy? Or with our own dead, who certainly seem to be involved, or—what is most likely—with the vast weave of conscious being that is described in the spiritual part of the Reddit document. No matter what, though, if this energy is conscious, we would be very foolish to assume that it is passive—or, for that matter, that our past methods of dealing with it through rituals like prayer and sacrifice are what it can relate to with all the depth and nuance of which it is probably capable. The reason I say this is that these practices do not bring a consistent response.

So how do we engage? In the final chapter, I will offer some observations about this.

The only siddhi that has been found to produce effective results using modern methodologies is remote viewing, a variant of clairvoyance, and these results are also inconsistent. There have been and are successful remote viewing attempts, but unless they are verified by follow-up confirmation, there is no way to tell whether they are imaginary. There is a great deal of remote viewing published nowadays that involves looking at structures on the Moon and Mars and such things, all of it unverifiable.

A famous remote viewing program, known as Project Stargate, was run for the Central Intelligence Agency at the Stanford Research Institute starting in 1972 and cancelled in 1995, allegedly due to unreliable results. Given that it had been running at that point for twenty-three years, there have been claims that the cancellation was because of political pressure, and also that the program disappeared into the black budget and still continues. Whatever the truth, there were useful results, at least initially, but, as in all cases of extraordinary states, no definite answers have been forthcoming about why the process might work, and over time, the results became inconsistent.

Three factors impede effective study of this energy. The first is the assumption that it involves a spirit world that our scientific community does not believe exists, so a negative bias is always present. The second is consistency and repeatability. The third, and the underlying cause of that inconsistency, is that we don't have a stable relationship with this force.

In the east, the sage Vyasa, author and narrator of the *Mahabharata*, was said to be able to both bilocate and levitate, as was the Buddha. Stories of levitation abounded in the Far East, most particularly India. Recent claims, though, have not been proven. In the west, levitation has been part of Catholic tradition at least since the medieval period, inspired, no doubt, by Jesus's famous Sea of Galilee excursion. Stories abounded in the medieval–early modern period, and have persisted into the present time.

During the twentieth century, some successful efforts were made to document the phenomenon. The most recent story, hardly rigorous in its methodology, that has come to my attention, is a delightful one involving English grammar school girls.

It appears in a letter in the *London Review of Books* of August 15, 2024. The letter is in response to Malcolm Gaskill's May 9 review of *They Flew* by Carlos M. N. Eire, which discusses the medieval–early modern accounts, and which I shall shortly address more fully.

The *London Review* correspondent writes that she and some school friends, when they were about fifteen, used to play a game that involved having one girl lie on the floor and another six sit around her, touching her with the tips of their fingers. They would then chant in turn, "She looks pale—she looks ill—she IS ill—she looks dead—she IS dead." With just their fingers supporting her, the girl would "levitate about a foot off the floor and remain suspended for a few seconds before settling gently back to the classroom floor." After a member of the staff allowed herself to be levitated, the girls made plans to sell levitations in a tent at the annual school fair, but the headmistress got word of it and banned the practice.

This is, of course, a familiar party trick, known as "Light as a Feather, Stiff as a Board." Having experienced it myself, I can attest to the fact that one does feel surprisingly light, but I can assure you that, if the fingers are withdrawn, you don't "settle gently back" at all. My recollection is that the chant is "light as a feather, stiff as a board," not something about being dead. The correspondent's method would seem to be at variance with the common practice, a subtle difference which may have significance.

If the subject of the girls' variant was not guided down, but rather let go and floated down, then perhaps the letter reflects a little something more than what usually happened when I was a boy, which involved solemn promises not to drop the subject that were, of course, rarely honored.

To be specific, could the suggestion to the subject that she was "dead" have induced a change profound enough to have reduced both girl and clothing to the few ounces of weight that would be necessary to insure a gentle settle to the floor? It sounds impossible, of course, and the correspondent is unclear.

The book that inspired the letter, *They Flew*, concerns the tradition of saintly levitation that spread across Catholic Europe between 1300 and about 1700, when it gradually faded away. The book doesn't cover the modern era, but there have been many documented cases which we shall shortly explore.

An Italian saint, Padre Pio (1887–1968), might have been the last Catholic levitator to have been witnessed while in float. The claims, of course, are controversial, but it remains true that his fellow monks say that they saw him levitate during moments of spiritual ecstasy, and, as we shall see, this state was characteristically present when the levitators discussed in *They Flew* took to the skies—or at least, to the rafters.

Padre Pio was reputed to have appeared above his town of San Giovanni Rotondo and caused bombers of the 15th Air Force, which was commanded by General Nathan Twining, to turn back in confusion when they saw a monk standing in the sky. However, the issue of whether or not San Giovanni Rotondo was ever targeted remains an open one. The town does not appear on the 15th Air Force's mission list for the time it was stationed in Italy, but such lists are not precise, because they record only official targets, not where the planes actually dropped their bombs, and it was not unusual in that era for the wrong target to be bombed.

One thing that mitigates in favor of Padre Pio's levitation is the fact that it was consistently reported to happen when he was in a state of religious ecstasy.

In *They Flew*, Professor Eire mentions no fewer than 39 levitators, many of whom were scrutinized by the Inquisition, which would attempt to determine whether the claims were justified and, if so, if they were of divine or demonic origin. You didn't want the Inquisition to decide that you were being lifted by demons, or that you were a fraud, as these determinations carried with them dire consequences. Of the cases covered by Professor Eire, three are singled out as having been declared frauds: those of Magdalena de la Cruz, María de la Visitación and Luisa de la Ascensión. The first two fraudsters were confined for life in convents not their own. The fate of the third is unclear. All the levitators were extreme religious ecstatics. This means that, when they entered the state of ecstasy, they entirely lost touch with themselves. Their egos disappeared. Or, to put it another way, they died to the world and joined completely to their god.

It is easy to look back into the past and assume that the witnesses were credulous, superstitious, and ignorant. By our standards, they were indeed superstitious and ignorant, but they were not fools, and the records of the inquisitorial investigations reveal exhaustive efforts to get at the truth.

In some cases, the investigators witnessed the levitations firsthand. Domingo Bañez, who was investigating Theresa of Avila, was in a congregation when he saw her rise up after taking communion. She grabbed onto a rail, and hung there, begging God to stop the levitations.

Joseph of Cupertino was, by all accounts, the most prolific levitator of the period. He was seen to rise by thousands of people, including Pope Urban VIII. When Joseph, who was overawed to meet the pope, bent down to kiss his feet, he instead shot up into the rafters before the pope and the large assembly of clerics that were present for the audience. He floated in ecstasy until he was ordered by his superiors to come down. The pope was impressed. Eire reports he told the Father General of the Conventual Franciscans that, if Joseph died while he was still pope, he would testify as an eyewitness to his levitations. The pope died in 1644, Joseph in 1663, so he did not get the chance.

But was it really a miracle at all, or is what we are looking at a very real human power that, along with the others being discussed in this book, is actually normal to us, just as it is to our visitors? If so, then it must be mediated by a physics that we have either forgotten or, more likely, never really understood. This physics must be somehow engaged by extreme mental states. Might it therefore be possible to gain enough understanding of it to enable us to harness the powers it describes in the same way we have harnessed so many other natural phenomena, from fire to the wheel to electronics? Put another way, what was that electricity that Paul Eno experienced when he touched the levitating girl?

Paul did not say that the girl was in a state of religious ecstasy when she levitated. Then again, she was from a Catholic family so devout that they had called on an exorcist to try to cure her schizophrenia by driving a demon out of her body, so she was certainly in an intense religious situation when the levitation took place.

Witnesses reported that Theresa of Avila, Joseph of Cupertino, and Maria of Agreda, who was also a noteworthy bilocator, all seemed to be dead when they were levitating. One wonders if Catholic folk-knowledge of that fact inspired the unusual "dead" chant in the "Light as a Feather, Stiff as a Board" party trick as the girls at the Catholic grammar school played it? If so, perhaps their subjects really did float gently to the floor, as they were unintentionally and in a small way replicating the state that the great levitators entered when they went into trance. Is becoming dead to the world a prerequisite for levitation? It does seem that, to activate siddhis, the ego must be profoundly controlled. In terms of the way the brain forms ego, once again, the default mode plays a central role, and there are studies showing that intense meditation and the use of certain psychedelic substances leads to the paralysis of the network, and with it an anesthesia of the sense of self.

Thus, it can be said that extremely deep states of mediation work like religious ecstasy to dissolve the ego.

But meditators, in general, cannot levitate, or, if they can, it is nowhere near as intense as what would happen to Joseph. His state would be so intense that he would become completely rigid, and so insensitive to physical stimulation that his feet could be pricked with needles, his hands put into fire, and his eyeballs actually poked with the finger, and there would be no reaction.

Maria of Agreda's body became so light that her sister nuns could blow it around with a puff of breath. She also was tortured while in this state, to no effect. In fact, she was so famous for her levitations that floods of visitors came to her convent, and her sisters took to floating her out into public where people could move her about by blowing on her as if she was, as Professor Eire puts it, "a wispy plaything." When she found out that she was being used in this way, as an amusement, and a paid one at that, she was furious.

I don't think that it's rational to dismiss all of this witness testimony and rigorous investigation, spread across hundreds of years, simply because we do not at present understand the biology and the physics that must be behind it. There would appear to be a state of some sort that causes the body and the clothing it is wearing to become weightless, while rendering the levitator catatonic. This state may be triggered by religious ecstasy or other ways of dissolving the ego, and possibly, in a small way, even by means as mundane as a common party trick.

I think that it's worth further exploring the possibility that the energy behind it might also account for other extreme distortions of normal physical laws, such as bilocation, out-of-body travel, healing, instantaneous movement, invisibility, imperviousness to heat and cold, and perhaps other unguessed, or forgotten, abilities which defy ordinary physics. These effects originate with the same state of body and mind—a state which our visitors possess and can control, but which is asleep in us. The reason we have not discovered it is simple enough: It's Paul Eno's electricity. We don't know what that is. But it is there, within our bodies, waiting to be measured, tested and understood.

What seems to be missing in modern times are two things: the first is intensity; the second, the belief that these extreme states are possible.

If the ability to change our physical state as completely as is implied by levitation wasn't present in us at all, these witness accounts wouldn't exist. But they do. Professor Eire never advocates in his book for belief, but the vast amount of testimony he has gathered speaks for itself. Something we do not understand happened to these people, and the case that it caused them to levitate seems strong.

As must be clear by now, I am not a believer in the supernatural, but in Jeff Kripal's concept of the 'super' natural, that is to say, natural powers that transcend the physical as we perceive it, but are still very much part of nature. Starting in the Renaissance and intensifying with the development of the scientific method, we have increasingly rejected even the possibility that they exist. It appears, though, that they do. Therefore, they are part of nature and there are natural laws that govern them, which we can seek to understand and utilize, just as we have so many others.

There was an upsurge of interest in levitation during the spiritualist movement in 1848, when two sisters, Margaret and Kate Fox, claimed to communicate with spirits through knocking sounds. In 1888, Margaret Fox stated in a lecture that she had produced the rapping sounds by cracking her joints, but later recanted. Meanwhile, the movement spread, taking root in the U.S., England and throughout the Western world. One nineteenth-century practitioner, Daniel Dunglas Home, was observed on some occasions to levitate. Home was witnessed by many people, including Mark Twain, who believed him to be genuine, and a noted scientist of the period, Sir William Crookes, observed levitation and found no evidence of fraud. Of course, there were also many credible observers who suspected fraud, and understandably so, but none was ever proven.

In the twentieth century, efforts to document levitation continued, and photography provided some interesting evidence. In his thorough treatment of the subject in *Human Levitation,* Preston Dennett details a substantial number of these cases. One of them, originally reported by Charles Rodney and Anna Jordan in *Lighter than Air: Miracles of Human Fight from Christian Saints to Native American Spirits,* quotes from a letter from German veterinarian, P. Mueller, who was posted to Türkiye in 1916 during World War I. He was in the audience watching a group of Howling Dervishes (Rufai Dervishes) enact one of their intense dances, shaking and jerking and, presumably, howling, when one of them suddenly rose about eighteen inches off the floor and remained floating there for some time. As an aside here, I have noticed quite a few levitation sessions involve a movement to about that height, which is also roughly the height I once saw one of the visitors reach as they levitated in order to race away from me. I don't know if this will prove to have any significance, but I want to mention it.

During the 1930s and 1940s, there were several successful efforts to photograph mediums while they were levitated. Danish photographer Sven Turck organized a group dedicated to levitating a human being and photographing the event. A medium called Borg Michaelson rose to the ceiling, circled the room, and then dropped to the floor. The room was dark, but not so dark that the levitation was not recorded by three cameras. They show Michaelson in a semi-crouched position, fully ascended, in different positions above the heads of the séance sitters.

I have personally witnessed the levitation of "spirit trumpets" during the weekly private séance of British medium Stewart Alexander. One of them lifted off the floor and came right up to my face. When I thought to myself, "This must be a trick," it proceeded to rub itself against my nose, as if to rub my nose in my doubts!

I examined the room and the trumpet both before and after the séance, and I could see it easily in the dim red light that was turned on during the session. It was not connected to wires, and, as it was made of paper, could not have been manipulated by magnets. There would have been no point to trickery in that séance. The reason is that Stewart conducts them privately. No money is involved, and he has been sitting with the same small group of people for years. Outsiders are rare in the séance. What would be the motive for a group of old friends to sit in a little room in the dark weekly for over forty years to witness the same trickery again and again? Given what I saw, it would have to involve extraordinary effects, and what would be the point of carrying out such masterful tricks over and over for just a few friends? I would assert that, in fact, Stewart's séance doesn't involve trickery. There is no motive.

I think it's clear that any open-minded study of the data must compel the conclusion that the levitation of both human beings and objects is possible and does happen. The rejection of this reality results from a profound failure of culture, intellect and imagination, and is symptomatic of what is preventing us from recovering our lost powers.

Why are we like this, though? One would think that such a remarkable thing as human levitation would be the subject of much study. After all, whatever it is, it causes the human body and the clothes that cover it to become lighter than air. There is absolutely no known force in nature that could do this. Not only that, animals are not seen to levitate, only us.

Only us ... and our visitors.

It seems clear to me that something can cause actual, physical changes in the atomic structure of the cells by an unknown mental state that has something to do with ecstasy and something to do with concentration and trance.

Because it causes bodies to become so weightless that, as in the case of Mary of Agreda, they can be moved through the air by simply blowing on them, it does not involve unseen spirits lifting them. Invisible presences may well be involved, but something is happening to the bodies at the atomic level or even deeper than that. What, we do not know, but if we are ever to recover what we have lost, we need to make a serious and rigorous effort to find that out.

It is not in the scope of this book to offer solutions to this mystery. Instead, I would ask: If we once had this power—and others—why were they lost? And that, I think, can be productively explored.

To do this, we need to travel back in time about 40,000 years, and then to a more recent period which began 12,900 years ago, and which, over the next 2,000 excruciating years, almost brought the human journey to an end.

CHAPTER THIRTEEN

THE TERROR THAT CAME IN THE NIGHT

What kind of shock could literally destroy a mind, leaving it a shadow of what it had formerly been? Here, we are looking at trauma that would have continued over an appalling forty human generations. This would mean that by fifty years into the catastrophe, nobody alive would have known a time when the skies did not rain fire, wild storms did not blow up out of nowhere, the oceans did not keep rising and tidal waves come gushing ever further into the drowning lowlands of the world—where something that we might call a civilization may well have been struggling for its life.

The period of upheaval started about 12,000 years ago. It was named after a cold-weather plant called the Dryas, whose range expands when planetary temperatures drop. The Older Dryas took place about 2,000 years previously and was not as intense. We know it occurred because it saw a brief expansion of the range of the Dryas. The Dryases were dated, in part, by measuring the length of the Dryas expansions.

But what caused this upheaval?

As it involved impacts, fires and floods, there are few possibilities other than a tremendous increase in debris striking the planet. While a debris field, roaring through space, could be caused by several things, the most likely one is a supernova, specifically, the physical debris from such an event. If this happened, there would be some sign in the distant past of the passage of the light and radiation from the explosion, which would move much faster than the physical debris.

There was such an event 30,000–40,000 years ago, which is signaled, among other things, by a minor extinction that involved both Southeast Asian and Australian megafauna. As these animals would have been continuously exposed to the sky, it is possible that they were decimated by gamma rays and other forms of radiation that would have been emitted, along with visible light, by a supernova. If so, this would mean that the greater part of the supernova's radiation must have struck that region, which is, surprisingly, supported by a legend from Taiwan, in the same area. There is also some evidence of a decline in the human population in Australia during this period.

What people living in that more distant time and in the right area of the planet would have seen would have been a second sun coming out of nowhere, fiercely bright, transforming the night into a malignant parody of the day. The birds would have woken up, flooding this ghastly new day with song. Night-blooming flowers would have closed, drawing away into themselves. The beasts of the night would have scurried back into their burrows and dens, and the beasts of the day ventured forth to forage. People would have awakened, too, and looked up, and been afraid.

What may be a memory of this, in the form of a legend, has been handed down for countless generations by the Atayal people of Taiwan. It tells of a time when a second sun did appear in the skies. The story of the "twin suns" says that a second sun began to rise when the first one set, causing extreme overheating and drought.

There is very little the story could be about except the appearance of a supernova in the skies of Earth, and indeed, we are in a region of our galaxy where this specific type of stellar explosion is more likely to take place. This is because we have, for the past few million years, been passing through a galactic arm. This is visible in the night sky as the great arch of stars we call the Milky Way. Because of its density, there is a greater chance of a supernova here than in less populated parts of the galaxy, and indeed, the specific area we are in, known as the Local Bubble, has been formed over millions of years by supernova explosions.

But still—a legend that has lasted in tribal memory for 40,000 years? We don't even possess genetic records of the Atayal that go back that far. But I refuse to colonize their story with Western assumptions that argue that such a long-lived folk transmission is impossible, so I'm simply going to leave it at this: The only supernova powerful enough to cause the effect that is described in the story appeared about 40,000 years ago, and, judging from megafauna extinctions it would have been at its most powerful in Southeast Asia, which is where the Atayal people live.

There are more very old stories, some of them much older. One, in particular, must be ancient indeed. It's not a story, really, but the name of one. It is the fact that the Pleiades are called "the Seven Sisters." The problem is that only six stars are visible. This is not a recent development. Aratus of Soli (315–240 BCE) mentions in his long poem *Phenomena* that there are only six stars visible in the Seven Sisters, meaning that this anomaly was noticed all that time ago. Cultures around the world record the reason for the missing seventh star with different legends, but all agree that there are only six stars visible to the naked eye of the average person. The last time there were seven stars visible in the constellation was, at the latest, 60,000 years ago. According to a paper by Ray and Barnaby Norris, "Why are there Seven Sisters?," two stars, Pleione and Atlas, once appeared further apart. The authors argue the constellation would last have been visible as seven distinct stars no later than 60,000 years ago (Norris and Norris 2020). This would mean that the number of stars in the many stories about the Seven Sisters has remained unchanged all that time, and therefore that the story has come down to us from the middle Pleistocene. *Homo sapiens* were already present in much of Africa, Asia and Australia and southern Europe, and Neanderthals were not yet extinct. It seems probable from this and other very ancient stories that they can remain intact for an extremely long time, giving additional credibility to the Atayal tradition.

Besides the extinction of the megafauna in Southeast Asia 30,000 years ago, there was a reversal of the Earth's magnetic field known as the Laschamp excursion. Such an event, falling outside of the broader cycle of reversals, could have been triggered by extreme stress on the field of exactly the sort that would be caused by a supernova. In addition, ice cores, marine deposits and tree ring evidence show a spike in carbon-14 and beryllium-10 isotopes during the same period. This means that cosmic radiation reaching Earth also increased, another possible marker of a supernova.

The question then becomes: Can we identify the supernova that was involved?

John Ellis and colleagues argue in the *Astrophysical Journal* that the Geminga pulsar, which is now 800 light-years away, is the remnant of a supernova that exploded between 50,000 and 40,000 years ago and could have been 100 to 300 light-years away from Earth at the time (Ellis et al. 1996). If a supernova took place then, its physical debris would have reached Earth between 10,000 and 20,000 years ago, depending on velocity. This is what would have been responsible for the Dryas events.

But there is another possibility. There are, across a number of major world cultures, stories of a war in space, or a war between the gods, that had a devastating effect on Earth. In the past, I would have dismissed such stories, but I feel now that the uncritical imposition of modern Western ideas about the past on indigenous accounts has caused them to be too easily dismissed. As we shall see, whether these accounts are actually about space wars or attempts to explain cosmic events that could not be comprehended any other way when they happened, respecting them will lead to some powerful insights about why we appear to be such a deeply traumatized species.

In one way or another, stories from many of our oldest cultures record wars near and on Earth between technologically advanced and aggressive presences, usually described as 'gods.' Such stories come from ten large ancient cultures and many less extensive ones, especially among indigenous populations. Most of these stories, while they often involve battles between heavenly deities, are essentially creation myths, as the battles, such as the earliest surviving story of one, the Egyptian myth of the conflict between Set and Osiris, describe a struggle between chaos and order, with order winning in the end.

To me, this suggests a memory not so much of an actual war in space as of a period of natural chaos that eventually resolves itself—very much as the Younger Dryas did—although not before a thousand years of hell on Earth.

We struggle to imagine what that must have been like. To them, it would have seemed like forever—as if the world had become something other than what it had been, and the reason for the change would have been inexplicable.

Two stories, both dating from much later than Osiris and Set—the Hindu Mahabharata and Ramayana—involve detailed descriptions of what would have been a war in space. In these stories, warcraft called vimanas are described that could travel at great speed, ascend vertically, cover interplanetary distances and hover. Their propulsion systems are vaguely explained as being connected with mercury, but we have no knowledge of how such a thing might work. A more recent document, the *Vaimanika Shastra*, probably—and controversially—dating from the early twentieth century, explains the propulsion system in various ways, one of which involves a circular device that spins rapidly. I have been close enough to a modern UFO while it was slowing down to hear something that had been spinning inside it at high speed coming to a softly clattering stop.

The flight characteristics described in the two ancient texts also match those of the TicTac and Gimbal UFOs, as well as what was described in the September 23, 1947 Twining Memo, in which General Nathan Twining, then the commanding officer of the Air Materiel Command at Wright Field in Dayton, Ohio, describes their functionality as displaying extraordinary maneuverability and high speed. One difference is that the craft recovered from the Roswell Incident supposedly did not have a visible propulsion system, but it could be that, if the crash was intentional, it was removed beforehand to prevent us from examining it and learning its secrets. Maybe it had a propulsion system, as I suggested previously, that was so exotic that we simply didn't see it for what it was. Another explanation is that we found it, but that skilled social engineers in the intelligence community have created the impression that we didn't, in order to hide that fact.

But could the Younger Dryas really reflect the devastation caused by some sort of interplanetary war, one so great that it left humankind in a permanent state of shock and caused us to lose the abilities which we have been discussing, that are our birthright? Or, as certain mystics, most notably G. I. Gurdjieff, have claimed, were they taken from us, perhaps because we were on the losing side? In Gurdjieff's novel, an apocryphal history called *All and Everything: Beelzebub's Tales to His Grandson*, the main character, Beelzebub, tells his grandson Hassein that God, working through two archangels, placed something called the Organ Kundabuffer into humanity. According to Gurdjieff, this was done to reduce our access to cosmic energies that were believed to be too powerful for us to handle, but instead the helpless state in which we were left led to what Gurdjieff characterizes as the rise of egotism and other negative traits. This story is unique to Gurdjieff's work, but contains subtle echoes of the much older dialogue between Hermes and his son Tat that appears in the *Corpus Hermeticum* (second to third century CE). The stories are structurally identical, insofar as the older character imparts wisdom to the younger one, who serves as a stand-in for the reader. Both stories involve cosmic upheavals of sorts as well.

They also explore the idea that we have been somehow made less by something that happened to us in the past.

It's easy to see the concept of "kundalini buffer" in the word "kundabuffer" and think that hidden in this mystic's elaborate cosmology, perhaps drawn, at least in part, from unknown ancient sources and to an extent from the *Corpus Hermeticum*, is the idea that we were cut off from a deep relationship with some sort of cosmic energy, from which we drew the powers that we now no longer possess.

The Hindu tradition of kundalini describes it as an energy that is coiled at the base of the spine and which, with persistent effort at transcending ego and surrendering to higher powers, may uncoil and rise up the spine. As I have experienced this phenomenon, I do feel that it is real. The rising of kundalini is a fearsome thing, the unfolding of a ferocious and yet incredibly sweet and poignant sort of vibration in one's own body. And, like what we saw with the levitators, it seems to occur when the sense of personal identity—the ego—becomes just another part of one's being rather than being perceived as all that one is. Just as what happens when psychedelics shut down the parts of the brain that support ego, a completely different vision of the world, which is one that does not rely on our ordinary expectations and assumptions, is revealed. Kundalini feels like electricity, sometimes mild, other times intense.

If there was a war, did our loss of it cause us to retreat or be forced into what amounts to a false reality, governed by a self that doesn't really exist, but which so diminishes us we can only achieve very muted recoveries of our lost powers through extreme deprivation and ego destruction, the sort of things that the medieval religious ecstatics visited on themselves?

With the exception of the two Hindu tales, which are among the more recent of these stories, the myths look more like creation tales and anthropomorphized records of cosmic disturbances that fit the natural upheavals of the Younger Dryas and its eventual end better than they do wars. But whomever was able to do things like build Baalbek had exceptional powers, and might have also fought wars in ways that we no longer understand. If this is the case, the debris left by their battles might look like the remains of natural catastrophes, or it could even be true that the pressures of the Younger Dryas led to conflict, and what we see now is a combination of ruins caused by natural catastrophe and ruins caused by war.

And yet, if we are the devastated and captive remnant of a once-cosmic species that lost a war in space, it would be in our captors' interest to prevent us from knowing the truth. For if we did, would we not set about trying to free ourselves, possibly by doing exactly the sort of things that this book proposes?

Of all the thousands of stories of contact that I have been told, one of the most disturbing came in a letter from a woman in Tennessee, who said that she'd been having a walk in the woods with her son when a strange-looking little man came out of a cave and told her he was a rebel and he had a message. It was that there had been, in ages long past, a war between Earth and Mars, and while Mars had been stripped of all surface life, they had captured our souls, and have condemned us to return to physical life after death, and never ascend into higher realms. Their name for Earth is "Dead Forever."

Just a fanciful tale, or a genuine message? I wish I could say that I knew.

What I can do is describe what we do know happened all those years ago, and how it must have affected people living then, and why, if there were large populations and even large population centers, they may have disappeared entirely, as alternative archaeologists like Graham Hancock assert. His contention is that there has been so little underwater archaeology that no useful conclusions can be drawn about whether there were greater population centers prior to the appearance of the earliest civilizations in India, Sumeria and Egypt around approximately 3000 BCE, which is what conventional archaeologists believe.

As Hancock discusses in his book *Underworld: The Mysterious Origins of Civilization*, underwater remains, among them those found at Cambay in the Indian Ocean, have been dated to about 9000 BCE, which is about the same time as Meltwater Pulse 1B, the second of the great outpourings of water from the melting glaciers, took place. As the ice age ended, the two great meltwater pulses and other smaller ones collectively raised sea levels by a catastrophic general level of 400 feet worldwide, which is certainly enough to submerge a great world civilization. It is also notable that the Göbekli Tepe site in southern Türkiye began to be covered over by its creators at about the same time, suggesting that they were under some sort of unusual stress.

As there has been just the most minimal underwater archaeology ever done, and this sea level rise was so dramatic, it is inappropriate to argue that there could not have been more extensive human organization on Earth before the Younger Dryas, which we know caused incredible destruction. The claim that "no evidence has been found" of earlier advanced human activity seems to me as insupportable as the claim that aliens could not get here from other worlds because "the distances are too far." Such assertions may or may not be true, but, like their opposites, they are speculations, not assertions of fact.

What would it have been like to actually endure the Younger Dryas? It was without question the greatest catastrophe that the human race ever experienced. Also, it continued not just for a short time, but for well over a thousand years. Generation after generation knew nothing but a raging sky, tortured land and unending struggle and terror.

They would certainly have looked for a reason, so it's not surprising that we have legends of warring deities reaching us from the deep past.

The reason that the celestial events that characterized the period would have looked to the peoples of that time as if a war was going on over their heads is that the debris involved—which would have included dust, larger chunks of solid material, and probably cometary material captured by the incoming wave as it passed through the Oort Cloud—would have set up a spectacular and generations-long period of meteor showers, asteroid strikes and cometary passages, some of them undoubtedly involving violent impacts.

It is almost impossible to conceive how it would have been to live through such a time. It lasted so long that the memory of when it started would have been lost in the generations. It would have been, simply, always—the world as it is, a permanent reality. Impacts striking the sun would have deranged the solar cycle, causing severe flaring and coronal mass ejections. These would have resulted in fantastic auroras that would have filled the night sky with blazing forms that would have looked to those people like some sort of living fire. Pareidolia, or the natural tendency of the brain to conflate ambiguous shapes into apparent faces, would at times have made these discharges seem like gigantic beings raging through the skies.

The Younger Dryas started suddenly, with a catastrophic impact that had an immediate effect on North America and resulted in worldwide climate upheavals and flooding. It did not start with flooding, though, but with fire.

Around 12,900 years ago, North America caught fire. This must have been caused by a rain of meteors large enough to reach the ground, and to have done so over a wide area of the continent. Most of the continent was densely forested, with more open plains closer to the Laurentide Ice Sheet, which at that time penetrated as far south as Montana in the west, Kansas in the central U.S., and New Jersey in the northeast.

Immediately south of the glacier there were outwash plains, which were caused by meltwater streams. They abounded with wildlife, especially megafauna like woolly mammoths—aggressively hunted by the Clovis people, whose archaeological remains suggested they were the dominant human population at the time. They produced distinctive bifacial flint arrowheads known as Clovis points, which have also been found in Mexico and Central America, as well as southern France and Spain.

The idea that the Clovis people might have had some connection with the Solutrean populations of France and Spain is not widely accepted, but, as will be seen, there could be a surprising explanation for the similar toolmaking that is found in these and other widely separated places.

Probably over a matter of just a few days, the world of the Clovis people burst into flames. As the first wave of debris entered the solar system, it would have affected the outer planets and been drawn into the Sun by its powerful gravity. This would have caused auroras so intense that they would have been visible day and night. To people who had no understanding of nature, this would have been terrifying. If there were more sophisticated cultures clustered along the coasts, they would have seen this as ominous.

There would have probably been visible events on the Moon, but we have little information to date impact craters precisely enough to know if any coincide with the Younger Dryas period. Given what we do know happened on Earth, it's likely that there were impacts on the Moon.

Large lunar impacts can be seen with the naked eye. In fact, the Giordano Bruno crater was probably formed by an impact that was seen by monks in 1178; the brightest flash ever recorded on the Moon took place in 2013, and could be seen by the naked eye. In any case, judging from a spike in ammonium levels found in Greenland ice cores dating from 12,340 years ago, it can be inferred that the great North American fires were taking place at that time.

While the first wave of the physical debris field began moving through our solar system 16,000 years ago, it was not until over 3,000 years later that the situation became catastrophic. The first thing that the Clovis people would have become aware of was a horrific bombardment of high-speed micrometeorites. These would have penetrated millions of animals to the bone, and probably many of the Clovis people as well. Fossils of most of the large animals of the period can be found with one side of tusks and bones impregnated with tiny particles of magnetic iron dust.

This would have caused an agonizing but almost immediate death, and could explain why some mammoth carcasses are found with food still in their mouths.

There was a fantastic group of impacts on the Laurentide glacier, one of which splashed multi-ton fragments across the continent. These struck the ground in the Carolinas, forming what is now known as the Carolina Bays. Others were hurled across the continent, some impacting as far away as New Mexico.

Many more objects would have struck the oceans, meaning that coastal and island communities would have been subjected to tsunamis and tidal flooding, and all of this would have happened, for the most part, without warning. It is possible, though, that people with an understanding of the skies would have been concerned about what was happening. They might have seen new stars that were increasing in size as the material approached Earth, but they also would have been mystified by the initial high-speed spray of micrometeors. Their communities would have been swiftly inundated by the tidal catastrophe. The ruins, likely severely damaged, would, over the next few years, have disappeared forever beneath the rising sea waters.

There would be survivors, of course, all over the world except in North America, where only the tribes who lived along the Pacific coast from Oregon and into Canada survived.

Everyone who survived would have been profoundly traumatized. The disaster would not have ended, though. Far from it, disruptions of varying intensity would have continued for another 1,400 years, during which time the world would have at first become dark and cold, then heated up again so dramatically that the already stressed glaciers would have disappeared into massive floods. At one point during this period, the Mississippi River would have been a roaring, impassable cataract, like a giant moving lake. Animal populations would have collapsed. Famine and elevated radiation levels in the atmosphere and soils would have weakened immune systems. Illness would have taken many, starvation even more.

Populations obviously collapsed almost completely in North America, and a 30 to 60 percent drop is estimated for the rest of the world. European and Asian populations hugged riverbanks and coastal areas, following the wildlife. About 9,000 years ago, or during the period when the Younger Dryas was ending, evidence of farming first appeared in the Levant, most notably around the ancient town of Jericho. But it may well have started earlier, albeit on a small scale, around coastal communities that have been lost to rising sea levels.

What did this catastrophe mean to the people who experienced it? What did it do to their minds, and what was lost as a result?

We turn now to what we lost of ourselves during that terrible time, and why we lost it, and what it might take to bring it back.

CHAPTER FOURTEEN

SHADOW OF A CATASTROPHE

Humankind spent nearly two thousand years in a kind of hell. The Younger Dryas was a catastrophe so great that we simply have not been able to grasp its full meaning, let alone understand its effect on our bodies, brains and minds. It involved such profound generalized destruction, including the redrawing of every coastline on the planet, that, if there were organized human communities present before it happened, they and all direct traces of them have disappeared.

Despite the extensive and undeniable evidence of the tremendous scale of the disaster, and a substantial number of difficult-to-explain artifacts and ruins around the planet, there has been no robust effort to map the pre–ice age coastlines in search of likely sites for ancient communities. Instead, what we have done is dig in what were in those days uplands, which are generally less well populated than regions closer to the oceans.

If there was a civilization, it was not an early version of what we have now. It did not function according to the technological norms that we have established as indicators of civilized society.

It probably didn't have cities as we now understand them, and the technology it utilized was much more like what we see our visitors using than what we have developed during the historical period. They may not even have had tools like the written word, but rather communicated in a different manner, more similar to what I and other close encounter witnesses have observed among the visitors, which is mind to mind telepathy.

As we shall see in the next chapter, if we are to gain a really useful understanding of who we are and why we are as we are, we have to make a much more objective effort to understand ourselves than we have done so far. Coming to a genuine and correct understanding of what "know thyself" means in terms of our origins is the single most essential key to unlocking a long-term future for ourselves.

With that aim in mind, I would like to explore an area which has not been studied from a paleoanthropological perspective. This is what protracted trauma did to brains and minds during the Younger Dryas.

To begin, let's review what we know about the effect that protracted, non-injurious trauma has on the mind. Obviously, many people would have experienced injury trauma, ranging from physical injuries of all kinds to famine and disease. But essentially everybody would have had the anguish of witnessing the deaths of children and friends, the desperate terror involved in fleeing from violent storms, floods and fires, and observing terrifying and bizarre auroral events. On the North American continent, the few survivors would have been clinging to life around the peripheries of the vast, toxic lake that was what remained after the continental fire and the flood that followed it. This lake stretched across the entire heart of the continent. When it evaporated, the ash and plankton shells that were left behind became the Black Mat, which is a layer that primarily consists of charcoal and the skeletons of algae. This material was deposited as a result of a massive, continental fire which was followed by an equally huge flood. This almost unimaginable event involved a vast area of the North American continent.

Even though the people living prior to the advent of the Younger Dryas were attuned to the energies of the planet in ways that enabled them to move gigantic stones and probably do many other remarkable things, it is unlikely that those abilities would have done anything to improve the situation. Their powers would have ceased to matter. Escaping environmental calamities and finding food would have preempted everything else, including even healing. Contact between minds over long distances would have been thrown into chaos by vast destruction and death. If these powers were not universal, but confined to limited number of people, most of them clustered in coastal communities, this would have been even more true. The sheer pressure of constant, immediate danger would have disrupted skills that require extreme concentration such as levitation. Most importantly, the ego would have expanded dramatically, an adaptation essential for survival of the individual. Extensive, even worldwide social groups connected by telepathy would have first contracted, then disappeared as individuals struggled to maintain local and individual survival. The larger human groupings, probably of a kind that we no longer know or even imagine could have once existed, would have disappeared into the kind of highly localized tribal units that we still see in non-literate and minimally literate populations.

This would not have happened only as a response to adverse conditions. It would have been forced by brain changes that are associated with protracted exposure to trauma.

The major changes that have been documented among modern major trauma survivors are as follows:

The hippocampus, which, as we have seen is essential to memory and learning, shrinks in size due to exposure to stress hormones like cortisol. In people with post-traumatic stress disorder, which would have included the entire Younger Dryas population, these changes in the hippocampus are responsible for incapacities like an inability to distinguish between past and present experiences and to form new memories. This condition isolates the individual in ways that trigger an anger response, because they cannot properly respond to the social and environmental inputs that they are experiencing. To somebody needing an extreme ability to concentrate to achieve states like levitation, the effect would have been devastating.

Disruptions in the brain's cortisol regulation would have sent the pituitary–adrenal axis, which controls cortisol release, into chaos. These disruptions would also have affected the pineal gland, which could be the core reason that we lost the powers that are under discussion here.

While modern science cannot detect any of the traditional aspects and powers of consciousness associated with the pineal, this could be because it no longer works as it should.

Modern neuroscience also records that stress causes disruptions in the pineal gland's ability to produce melatonin. It can induce rapid aging in the gland, causing it to become less effective. The apparent presence of magnetite in it might also have something to do with its capabilities, but this is not understood at this time.

Hindu tradition relates the third eye to the pineal gland and contemplates the third eye as a sort of transceiver that enables things like far-seeing and telepathic communication as part of its ability to see into higher levels of reality, and receive signals from what are thought of as subtle energies. Buddhism similarly has traditions that the third eye is related to higher forms of sight, and various forms of shamanism use substances like ayahuasca to generate DMT floods.

Are beliefs that the pineal has special powers a memory of a long-ago time when it did? We know that stories persist in the human experience for a very long time. Why not other memories that are similarly carried in tradition?

The people who were using extraordinary powers of mind prior to the inception of the Younger Dryas probably had no knowledge of how they were being produced. The powers of the gland would have been first disrupted, then lost altogether. The abilities it once conferred would have become distant memories, passed down in the more robust cultures until, when the chaos ended and we once again began to be able to create complex societies, we would probably have tried to replicate the processes it once enabled. As to it seeming unlikely that memories of lost powers could go back to more than 12,000 years ago, we have already seen that stories can persist for much longer than that.

One can scarcely imagine how it would have felt, during those long ages of disaster and chaos, to have lost the powers that are attributed in tradition to the pineal. Just when they were needed the most, nature itself, by visiting such terrible trauma on humankind, caused them to fade away.

The elaborate instruments and rituals of our earliest religions may have represented attempts to use magic to bring those powers back, and the instruments used in those rituals might have been modeled on those of long ago, now drowned and buried in the oceans of the world, that once had a role in activating our lost powers through ritual.

It is for this reason that the eyes on the cover of this book are naked. If the Reddit document is correct that the great black eyes that we have seen are actually coverings, then it is time for those coverings to reveal what has been hidden for so long, as we look back across the reaches of time and see ourselves for what we truly are, and, I hope, begin to awaken from the long sleep to which nature, in its random violence, condemned us.

This awakening was perhaps prophesied as long ago as 1988, when Raven Dana's contact occurred at our cabin. I have written about this many times, so I will only briefly summarize it here. Anne had gathered a number of people at the cabin who she felt might have contact, largely based on descriptions that they had sent us of their previous experiences, and the fact that they were still actively engaged. She had a number of successful sessions like this.

In this one, Raven Dana found herself late one night with one of the little gray people standing beside her bed. It took her hand and sent the type of energy that I have identified as kundalini coursing through her body. At the same moment, she saw what she identified as the left eye of Horus hanging on the wall at the foot of the bed. As we had no such object in the house, it can be inferred that it was there as part of the communication. As it was the left eye, it is associated with the god Thoth and the moon, and, in Egyptian myth, connected to the eye injury that Horus received in battle with his dark uncle Set being healed by Thoth. It is also associated with healing feminine energy, which is as deep a stream of cultural change as exists in our world today.

The Eye of Horus was there to complement the flow of kundalini energy and suggest the end of kundabuffer, which can be seen as, at last, the healing of the psychological injury we received from that long-ago stellar explosion and the agony it visited upon our planet, and on us.

What were we like then, though, before it happened? Where are the great cities, the sparkling hordes of people, the armies, the riches? Everywhere you look, our current civilization announces itself. It is present to the ends of the Earth and beyond, covering the planet with its vast output.

There's nothing like that from the past. If there was, even if submerged, surely we would see evidence of it. Therefore, it can be concluded either that it wasn't there, or that it was so completely destroyed and deeply submerged that we cannot see evidence of it. Has the destruction of our intrinsic powers also caused us to forget who we are and lose contact with our true history?

Let's take a closer look.

CHAPTER FIFTEEN

THE SINGING SANDS

Planet Earth is scattered with thousands of unexplained sites, many of which bear curious similarities to one another. The engineering of most of them defies explanation. There are dozens that involve immense stone blocks, some so enormous that we would find it essentially impossible to move them with modern equipment. There are worked artifacts such as diorite jars found in Egypt that cannot have been created without diamond drill bits, and which are machined to such perfection that handwork must be ruled out.

In addition, the engineering techniques used to create the megalithic structures were in use across the planet. As an example of this, there are extensive. exquisitely fitted stone walls in Peru and Japan that were constructed using what must have been the same technique, that somehow enabled the working of the stones as if they were a sort of clay. The Ahu platforms of Easter Island, on which the famous huge Moai figures are placed, probably used the same technique. The Japanese site lies near Shurijo Castle on Okinawa. In addition, gigantic stones, which could be moved today only with enormous difficulty, if at all, are found in the Baalbek platform in Lebanon and in the Osirion in Egypt. The stones were cut with extraordinary accuracy using tools of which we have no record. And then, of course, there is the pyramid complex at Giza.

We cannot accurately date any of these ruins. Instead, archaeologists date them by association. For example, the Sphinx is associated with Khafre (2558–2532 BCE) because the face carved on it resembles those seen on his statues and a pyramid attributed to him is nearby. Similarly, the Great Pyramid is attributed to Khufu (2589–2566 BCE) because a scratching of his name was found in the upper relieving chamber just below the King's Chamber by archaeologist Richard William Howard-Vyse in 1837. Some researchers have pointed out that the writing style of the Khufu graffiti is from a later era, and some entries in Howard-Vyse's personal journal suggest that he was involved in shady activities while doing his research, but Egyptologists, pointing out that the graffiti are similar to workmen's marks found at other sites, generally believe that they are genuine. Even so, there is no positive way to date the pyramid or any of the other sites, and they are all spectacularly designed and engineered. The Great Pyramid, for example, is aligned so precisely that it is difficult to imagine something so large being sited with such precision even today.

There are underground sites in Türkiye, Japan, India and other places that show evidence of an extraordinary ability to tunnel. The underground tunnel systems in Türkiye were occupied like cities, and could have housed tens of thousands of people. The chaos of the Younger Dryas would have included significant meteor falls, dangerous solar flares and intense storms, which might have driven people to large-scale shelters.

If there was a more populated coastal region in those days, the underground sites may be locations where coastal populations moved when their existing centers were flooded. They not only went far into the uplands, but also dug deep, probably in an effort to escape danger from above as well as local hunter-gatherer tribes, who were likely to have been themselves suffering from famine.

Another site in India, the Barabar Caves, are carved to an extremely high level of finish, and defy explanation. Why such enormous effort was put into these structures is unknown, as is their purpose.

In fact, our planet is strewn with so many unexplained sites and incredible structures that it can be argued that there was a worldwide civilization in the distant past, and the brilliantly designed and highly engineered ruins and inexplicable artifacts found in so many different places are the remains it left behind.

But it is assumed, instead, that all of these sites must be far more recent than the Younger Dryas, and that sites built to uniformly high and inexplicable standards in so many different places must have been constructed by widely separated cultures that had no knowledge of one another. In truth, though, there is no really certain way of dating any of them.

Possibly the most inexplicable site of all is located in the Caroline Islands of the South Pacific, and it may lead us to some completely unexpected clues that have been laid down in the twentieth century.

The site is called Nan Madol and consists of as many as 750,000 gigantic basalt logs, some of them reaching fifty tons in weight, that were placed where they were in order to create a group of artificial islands. They have been explained away by saying that they must have been moved by canoe. In fact, however, there is no record of any technology in use before the twentieth century that could have cut them, let alone transported them.

Nan Madol, all but forgotten in its isolated Pacific location, is one of the great structural and archaeological mysteries of all time.

The current consensus among archaeologists is that the site was built by the Pohnpeian culture during the Saudeleur Dynasty around 1100 CE. According to local tradition, two brothers, Olosipa and Olosopa, are said to have constructed it by levitating the stones. They were said to be magicians who had come from a faraway place, and used their magic to lift the stones into place—and, presumably, also to quarry them. Their purpose was to create a fortified city from which they could rule the Pohnpeian people. This seems to have been effective, as the Saudeleur Dynasty lasted over five hundred years until it was overthrown in 1628. It is possible, of course, that Nan Madol dates from an earlier time. It is sited in a shallow lagoon that would have been far above the waterline during the ice age. In fact, it would have been on a hill. The nearby quarry where the basalt logs were cut would have been on a hillside. If it is that old, then it would have been a hill fort.

For a reason that will become obvious shortly, I am not going to argue that the story of the magicians cannot be true, and that the islands must have been constructed long ago. It would appear that some of the ancient knowledge involving the moving of massive stones was known in the United States in modern times, and therefore could also have been known to a few people in earlier times. I say a few for the simple reason that, for the most part, the great megaliths are also the oldest structures in the world.

Like levitation, there is simply no way that huge stones can be moved with some vague something called "magic," but, as in the cases of the other ruins mentioned here, there is also no other way they could have been moved. Of course, it's conceivable that, with great effort, a few fifty-ton blocks could have been quarried and moved from the quarry to the Nan Madol site on huge, purpose-built barges. But not a thousand or ten thousand, much less 750,000! This, quite simply, is not a possible thing to accomplish in the few years it apparently took to build Nan Madol.

The first Western account of the site was published by a British trader called Captain E. H. Lamont in 1867. When he found it in the early 1850s, it was already in ruins, and had probably been abandoned when the Saudeleur dynasty fell. The 750,000 blocks had to have been quarried and moved during the dynastic era—by a population estimated to have been no more than 35,000 people.

Put simply, the only explanation that makes sense is that some technique that is now unknown must have been used to construct Nan Madol.

There is a siddhi that relates to this—as well as to the medieval levitations. "Laghima" is not only the power to levitate the body, but also to make objects levitate. In *They Flew*, there is a story about the master levitator Joseph of Cupertino making a huge cross, too heavy for ten men to lift, turn "as light as a straw." Joseph, who was at that moment himself in a state of levitation, flew to it and moved it through the air with him. The story sounds fantastic and completely impossible, but so do levitation, telepathy, instantaneous movement, healing and all the rest of it. We must somehow extricate ourselves from the trap of assuming that natural forces we don't understand don't exist. The effort made by the Varginha alien must be taken to heart.

One wonders if the mysterious brothers who are believed to have built Nan Madol had any connection to the levitators who were flying in the medieval period. Unfortunately, they are identified only as magicians who came, according to legend, "from the west." Asia is to the west of the Carolinas, and both Hindu and Buddhist traditions speak of accomplished levitators, but there is no way to trace any connection.

We are left with the question not only of who they were, but also how they might have accomplished the feat. As it happens, there may be a modern story that will bring us closer to some understanding.

Someone who apparently possessed the Laghima siddhi was Ed Leedskalnin, who created the Coral Castle in Florida between 1923 and 1956. Ed worked alone, possessed no heavy equipment, and yet was able to move stones weighing as much as 30 tons despite the fact that he was himself a lightly built man under six feet tall. In the twenty-eight years from 1923 to 1951 that it took him to create the castle, there was never any evidence of any heavy equipment use. Had cranes, tractors, steam shovels and such been in use, people would have heard them and seen them on the site.

After Ed's death, simple handmade tools were found on the site that would have been used to carve some of the more intricate shapes into the limestone he used, but nothing that could have even remotely been used to move the stones.

Ed claimed that he had learned an ancient Egyptian secret that enabled him to accomplish this, but he never disclosed it to anyone. Considering all the marvelous ways such a power might have helped mankind, it is unfortunate that he made this choice.

The same siddhi that the medieval levitators used, or rather that became active during their episodes of religious ecstasy, would have been responsible for the construction of Nan Madol if it was indeed built in the Middle Ages, and for Ed's strange skill. As he was not a religious fanatic, nor, one might suppose, were the mysterious brothers who built Nan Madol, it can be speculated that extreme ecstasy is not the only state needed to cause this siddhi to become active.

Whether or not Ed personally levitated in the depths of the night will never be known, but he seems to have done this with his stones, and it is likely that he also quarried them in some unusual way. He obviously used limestone because it is soft and easy to work, but the absence of any saws or drills at his quarry has left his stonecutting method as much of a mystery as his method of moving the blocks.

I might note here that at both the Osirion and Baalbek, there is evidence that powerful, extremely accurate saws were used, but Ed's stonecutting was not done in that way.

Another thing that is unknown is how he kept his movements secret from the many people who took an interest in what he was doing. Could it be that he used yet other siddhis—*vashitva*, or mind control, and telepathy? As I have experienced both in use, including at the same time, I am not in a position to doubt that they exist. I once witnessed a very strange-looking person loading shopping bags with smoking materials in our local drugstore when Anne and I lived in San Antonio. While this was happening, the clerks and all the other customers stood as still as stones, staring straight ahead. I will never forget the look of sinister complicity that the man gave me as he left. As soon as he was out the door, everybody came back to life, none of them showing the slightest awareness that they had been "turned off." This is only one of many incidents of mind control I have witnessed, so I would not be surprised if Ed was able to keep his work from prying eyes using this much-doubted but very real siddhi.

As to where he might have learned such things, it is not the purpose of this book to explore a relationship between the siddhis and ancient Egyptian magic. It seems plausible to speculate, however, that there was one. As mentioned above, on the west wall of the entryway to the Pyramid of Unas, there occurs a description of the way the body delivers the energy of life experience into the spine, which is thought of as a serpent of light. This is done with the seven smaller serpents called ta-ntr, which encircle the spine. These seven ta-ntr correspond to the seven chakras, and are fundamental to both Egyptian and Indian ideas about body energy.

Was Ed thus telling the truth when he said he had learned Egyptian magic, and is it related to Indian magic, as would appear to be the case? If so, then it's possible that the reason Ed was so obsessive about keeping his secrets was the same as that of the Indian practitioner of siddhis. It is, basically, that delivering them into the level of ego will cause them to disappear. Therefore, secrecy is essential to their continuing to function.

Judging from the immense blocks of stone that were moved to create the Osirion, and so much else that is impossible about Egyptian architectural engineering, they did have these powers, and as the most extraordinary Egyptian structures are for the most part also the oldest, they were able to use them more effectively early on than they were later. Just as is the case with Ed, no tools have been found anywhere in Egypt that could have cut the stones of the Osirion and many of the other sites.

If Ed learned Egyptian magic, it would probably have been in his native Latvia. But details about his early life are skimpy, beyond the fact that he was an emigrant and that the love of his life was a girl he called "sweet sixteen," to whom he dedicated his great work.

But why the secrecy? Certainly, it is a strong tradition when it comes to siddhis. It is said that they will fade if they are used to attract attention. I would think that the reason for this is that they don't work if they are ego-driven. Like the medieval ecstatics, who for the most part lost every trace of self before they levitated, the practitioner of siddhis must engage in serious inner work in order to detach their attention from ego and begin to truly live outside of their given identity. Otherwise, they don't have access to the power. If Ed was a magician, he would also have needed to keep his ego out of it if it was going to keep working. This would be the primary reason for his being so secretive. Had he, say, allowed the press to see him at work, he must have feared, or known, that he would lose his abilities.

As the modern world becomes more complex and social relations more intricate, inevitably, ego also becomes more complex. In a sense, the more we understand about the world, the less in touch we are with our inner selves and, it would appear, also our true capabilities. There is an energy involved, instinctively perceived—or perhaps known from some hidden source—by G. I. Gurdjieff when he described the organ that had cut it off as "kundabuffer." I think that "kundabuffer" was the terror of the situation itself, as it unfolded across the two thousand years of the Younger Dryas. The extraordinary power to create great monumental structures didn't matter in a flooding, starving and burning world. Abandoned, the powers atrophied. By the historical period, they were, for the most part, a distant memory. It is their energy, I think, that Ed tapped. He knew and could defeat the secret of what Gurdjieff had perceived. In his book, Gurdjieff identifies an individual as having cut it off, and while ancient sources as well as modern stories tell of an unimaginably distant war, I think that it is clear that it was a natural event that must have looked like a war among gods, as, for years and years, the heavens flashed with strange and terrible fires and rained destruction.

I think that the old powers of mind were lost or perhaps even intentionally abandoned by peoples worldwide who concluded that their use of them were angering the gods. And it wasn't just levitation that was involved, but also other siddhis, such as telepathy, as we shall see.

Nature broke us. It's that simple. And now there are predators and scavengers here seeking to feed off our helpless greatness, and a few saints, too, trying to help the sleeping beauty awaken.

While he kept his secret into death, Ed did leave a hint about what it might involve. In 1945 he published a short book on magnetism. It describes the force as a current like electricity, but modern science thinks of it as a field, not a current, that is not like electricity but is related to it. This relationship is described by electromagnetism theory. Ed's view is that it is much more fundamental to the structure of reality than we presently assume. His book describes various ways of constructing simple electromagnets and magnetizing metals, but toward the end, he describes magnetism as a "cosmic force" that holds the Earth and the Moon together and keeps them from colliding with one another. This has little to do with modern theories of gravity, but there has been some research into the relationship between magnetism and gravity. While there have been studies exploring how gravity and magnetic fields may interact in extreme conditions, such as near black holes, there isn't anything, at least in the public space, that discusses such a relationship at ordinary levels of intensity.

Some videos of UFOs, made at fairly close range and with high clarity, reveal lines of force around the objects that are likely visible in the digital medium because the objects are generating magnetic fields so intense that they affect the cameras. In addition, there are many stories of UFOs causing cars to suddenly stop when they come into close proximity with them. The father of a childhood friend had this experience during the Levelland, Texas UFO incident of November 2–3, 1957. He apparently did not report his experience, but numerous other motorists did, and most of them experienced their cars coming to a halt when the object drew near. Because they would start up again after the object departed, it can be inferred that the ignition systems were affected. Practically the only thing that would do this without causing lasting damage would be exposure to a very intense magnetic field.

While we do not see magnetic fields as being related to gravity, the presence of intense magnetic fields around UFOs which are defying gravity would suggest that such a relationship does exist. Given the fact that Ed didn't have any unusual tools, it is also probable that the mind can control it in some way, which would be Ed's secret.

What this all adds up to is that levitation, both of the body and of objects, is real, as are mind control and telepathy. These are all human capabilities that have been observed even in people like Ed and the woman Paul Eno saw levitate. It's not that we don't possess them, but that we don't realize it.

I think that a clearer understanding of magnetism and how it connects us to the Earth is probably crucial to regaining our lost powers in an objective and repeatable way—at least, some of them. But not all. There is another power that we lost and can, I feel sure, regain, and that is telepathy. The reason that I am so sure that we can regain it is that, when we are with the visitors, we fall very easily into using it. Therefore, we still possess it, but, like the rest of our ancient powers, it lies sleeping now and will continue to do so unless awakened.

I think that what is happening to us now could bring this about, but only if we understand our situation clearly and do what we can to turn the storm that is breaking around us right now, in the form of rapid climate change and the pressure of the appearance of the visitors. To achieve this, we must look deeper into ourselves than ego can take us. While states like religious ecstasy can free us from ego, there is a better way. The ecstatics were obsessed with their love of God, and I suspect that Ed's obsessive love of "Sweet Sixteen" might have had a somewhat similar effect on his personality, so that in fixating on his love of her, much as the ecstatics fixated on their love of Jesus, ego was put aside.

I doubt that we are going to restore our powers and recapture our destiny through this sort of obsession, though. We must understand, each of us in their own deepest presence, what we really mean and who we truly are. This will have the effect of containing ego. Once that is done—once we truly know ourselves—we will reconnect with these lost abilities and the energy on which they depend. Right now, I am sure it seems very abstract, but that will not always be the case. This is because there is about to be a confluence of two of the most powerful events in the history of the species: the collapse of the planet's ability to support us and the coming of the visitors.

The pressure the combination of these two events happening in the same time frame will put us under is going to either break us, as the horror of the Younger Dryas did the last time, or it is going to cause us to soar above where we are now, regaining our new powers and becoming a cosmic species. There is no third alternative. Here, now, in this time, we are going to either be born alive or dead.

Of all the powers we have lost, the one we need most is probably not levitation. It is, rather, the one that is most able to restore the fractured human community, which is telepathy.

As matters stand, if this other aspect of consciousness that we like to call "aliens" ever shows up physically, Western society is going to look first for understanding to the one group least able to provide useful direction: the scientists. The reason is that our scientific culture rejects the soul, and, as I have shown in this book and discussed in others, this denied aspect of the human being would appear to be what contact is primarily about. The physical aspect is secondary, and looking to their physical presence for understanding will only be useful up to a point. At present, science stops where the soul starts. This is not the fault of scientists, but rather of the fact that we have been unable to use the scientific method as it is now constituted to study nonphysical consciousness. We can't detect it with instruments. At least, not yet. As I have suggested, if the visitors are indeed primarily conscious energy, then they will need to make a decision at that level about how they will interact with us who are stuck in the physical, and with which ones of us.

The fact that Ed did successfully move stones by an unknown means and that he wrote a book on magnetism hints that it is in some way connected to the secret. Levitators become lighter than air, and that includes the clothes they are wearing. Gigantic objects can be levitated, which means that they also become extremely light. We do not understand the fundamental power that enables these things, but if we can ever accept that it exists, we might be able to make some progress toward recovering not only what the Varginha alien so poignantly regretted that we have lost—the ability to self-heal—but all the rest as well.

Looking down into the Osirion or walking into the Barabar Caves and listening to the eerily intimate echo of one's own breath, one feels almost able to touch the secret that Ed Leedskalnin somehow acquired. But then the desert wind blows and the sands shift, and the secret is borne away.

Or is it? Perhaps Ed didn't learn it from a mysterious sage at all, somehow still in possession of ancient knowledge, but discovered it within himself, as he was borne deep into the mystery of the mind by his own towering love for the girl he lost, where, in his anguish and his great desire to create a monument he deemed worthy of her, he stumbled on the secret of the singing sands.

CHAPTER SIXTEEN

WHEN WE WERE ONE

Long ago, when this world was a very different place, somebody built something astonishing in a desert highland above what is now the Mediterranean Sea. These builders had engineering skills beyond any that have been known on this planet since. What they created was a gigantic platform, the largest stones of which weigh 2 million pounds, or a thousand tons. We know where the stone was quarried, at a place not far from the platform. We know where the causeway ran upon which it and the other stones in the platform were carried. We do not know, however, how that was done—but Ed Leedskalnin probably did, as did whoever built Nan Madol. As to why the platform was being constructed, that is also unknown. But it is by no means the only inexplicable structure in this world. Far from it. In fact, there is another huge monolith in Yangshan, China that weighs 16,000 tons and also exhibits parallel striations that are strikingly similar to those produced by modern mining machinery. Such striations are also seen on the megalithic stones found at the Petra site in Jordan.

There are also 200 rose granite columns at Baalbek, each weighing between 60 and 100 tons, that were transported over 700 miles from a quarry in Aswan in Upper Egypt. No matter their route to the site, at some point they would have had to be transported over mountains, through passes that lie between 7,000 and 8,000 feet. They are very precisely carved and polished, and yet the only stone besides diamond that could do this is diorite, and no diorite tools are found at Aswan.

There are many broken columns on the site. Two hundred of them, many still intact, were used in the Temple of Jupiter which was built by the Romans on the gigantic megalithic base. Whether they were originally quarried for that temple or found by the Romans at the site is unknown. We do know that Roman cranes were capable of lifting 60 to 100 tons, so they could have been used to move the Baalbek columns once on site, but transporting them from Aswan, while possible in ancient times, would have required an extraordinary effort.

There is no other ruin anywhere in the Roman Empire that comes close to approaching the massive size of the Baalbek platform. Even now, transporting the blocks involved would require the construction of specialized cranes, which in turn would put such a strain on the subsurface of the area that the ground would have to be massively reinforced in order to insure their stability.

Between the quarry where the stones were originally obtained and the platform, there is no sign of any such foundation. In addition, there is an incline up to the platform site, meaning that the stones had to be lifted. We have no idea how any of this was done. The blocks might as well have been flown from the quarry to the site—which, one might suppose, would be no surprise to Ed Leedskalnin.

Where the builders came from or why they built the platform are unknowns, as is when, but there is evidence based on stone-cutting techniques that the larger stones were cut a very long time ago. No Optically Stimulated Luminescence has ever been done on the blocks. This process can date the times at which stones were first cut.

As to where the builders came from, there is some evidence of the existence of a much more sophisticated pre-agricultural population in the general area. First, the Göbekli Tepe ruin, now dated to 12,000 years ago, is a few hundred kilometers from the Levant where the platform is located. There is also something present there, as we shall see, that suggests that it was part of a worldwide system of some sort, now lost, that lasted into the Middle Ages.

The platform is far more sophisticated in construction than the Göbekli Tepe ruins¸ and whether or not there is any connection is unknown. There also exist in Türkiye the cave systems I briefly referred to previously, such as the Derinkuyu caves, which are man-made and could house thousands. In fact, they are usually described as an underground city, and pottery shards recently discovered deep in this system suggest that it may be, like Göbekli Tepe, as much as 12,000 years old—dating, in other words, to the beginning of the Younger Dryas.

Baalbek was abandoned very suddenly, so suddenly that the largest block of all, in our era known as Hajar al-Hibla, the Stone of the Pregnant Woman, was abandoned in the quarry while it was still incompletely cut. Even more extraordinary, another stone was located below the Pregnant Woman that weighs in at a fantastic 1,600 tons. And there could be other stones below it that have not been excavated. There is another quarry a short distance away that contains a partially quarried stone also weighing 1,600 tons. Parts of this stone were cut off by the Romans to use in their construction on the platform. All abandoned. But why?

As we have seen, the Younger Dryas began suddenly and violently, and caused upheavals worldwide. At that time, the Baalbek region, which had been temperate, would have experienced significant changes in climate, most notably between 12,800 and 12,900 years ago, which is when the Black Mat fires were taking place in North America. These fires were far larger than the enormous 2019–2020 Australian bush fires, which caused a measurable shift in climate due to the smoke haze they injected into the atmosphere. The changes that the Black Mat fires must have caused would have been far more extreme, resulting in the temperature declines worldwide that we do, in fact, see across the planet. They would have been accompanied by a dramatically darkening sky, which would have persisted for years. A similar phenomenon took place worldwide in the year 536 CE, when the world went dark for eighteen months. But the Black Mat fires burned for a hundred years. As an entire continent burned, the whole planet would have been plunged into a state of darkness or semi-darkness that would not have lifted for generations. Given this, the abandonment of the Baalbek constructions becomes more understandable.

The cold that would have accompanied the darkness might well have driven local populations in Türkiye to dig into the soft stone in Cappadocia to find shelter. Agriculture also developed at the time, beginning to appear in the region at about the same time that the Black Mat fires were raging halfway around the world.

Animal populations would have dropped dramatically as species established in the region died off or moved to areas better able to support them. Early attempts at agriculture would have been desperate affairs. Human populations would have collapsed worldwide, including in what was obviously a developed area in the Middle East. But the construction at Baalbek was far more complex than what was happening at Göbekli Tepe and in Cappadocia.

This leads one to speculate that the stones may have been carved by somebody with a more complex society, but who has disappeared entirely. This leads one to contemplate the possibility that they came from somewhere else. The most logical possibility would have been the Mediterranean coast of Egypt. This area would have spread far beyond the current Nile Delta, and would have been among the most fertile places on Earth. It would have been slowly disappearing due to gradual glacial melt prior to the Younger Dryas, then thrown into chaos by the same factors that were affecting the rest of the world at the time.

Could it be that Baalbek was intended as a refuge against rising waters, and when the waters stopped rising and daylight all but disappeared, the confused engineers abandoned the whole project—and with it, perhaps, also the secret to moving great stones that was only sporadically rediscovered, first by the Nan Madol builders, then by Ed?

There is also evidence that this vanished people were excellent seafarers and understood how to create colonies that would take root and survive over the long haul.

A good example of this involves the island of Cyprus, which was populated during the same period. According to a paper by Corey Bradshaw and colleagues titled "Demographic Models Predict End-Pleistocene Arrival and Rapid Eexpansion of Pre-agropastoralist Humans in Cyprus," a series of large-scale seaborne migrations took place to the island from the mainland, either

Türkiye or the Levant, during the unsettled period of the Younger Dryas (Bradshaw et al. 2024). These were not small, piecemeal traverses of the 90- to 120-kilometer overwater distances involved, but a group of three or four large-scale movements involving thousands of people. Thus, the island was not only populated intentionally, but also by a social group that understood that certain minimum numbers were needed to insure that the population would not become extinct after a few generations.

That all of these things happened at a time when agriculture was just beginning suggests that what we assume to have been a hunter-gatherer population was far more sophisticated than we have supposed. I think it's obvious that some sort of civilization existed and has been lost, and the evidence on the ground proves conclusively that they were master engineers. The fact that we have no record of them beyond their masterful monolithic works also proves, one would think, that their world was destroyed with great violence. And indeed, the first part of the Younger Dryas involved massive fires, but the second was even worse, as the glacial melt became an unstoppable cataract, and sea levels rose worldwide very quickly, probably under conditions of even more extreme violence, until they reached an appalling 410 feet higher than they were at the start of the period.

Were such an inundation to happen today, world civilization would be badly damaged, if not destroyed. Judging from the inland ruins that we do see—the relatively primitive ritual space at Göbekli Tepe, simple communities like Çatalhöyük and the caves—the upland dwellers must have been less advanced than those whose world was drowned by the rising seas. The Atlantis story is probably a distant memory of those times, with a diffuse civilization idealized into a single grand city.

Who knows what the society that created Baalbek and the other inexplicable ruins worldwide might have been like? Who can imagine? And as to why they engaged in such incredible building projects—there simply is no answer. But one thing seems clear: they understood far more than we have supposed, and, one might speculate, even knew reality in a different way than we do. The reason that we don't understand their works would seem clear: We also, and more importantly, don't understand their minds.

Their brains were our brains, this we know. Our brains, yes, but not our minds. However, a careful look into the even deeper past might reveal more of the ways in which they were different, and begin to tell us something about what the Varginha alien regretted that we have lost.

We have now seen that levitation and the ability to move stones by some means that seems to be somehow related are demonstrably real. But what of other siddhis? What about telepathy and, say, healing? The Varginha alien regretted that we no longer had the ability to heal ourselves. Is that, also, a sleeping human power? Is telepathy?

I have experienced telepathy with the visitors, and I'm not alone. Many close encounter witnesses would agree, and when it happens, it doesn't seem especially miraculous. It seems natural and easy and unsurprising. But then when they're gone, so is the telepathy.

As this ability exists within us, can we awaken it? And, for that matter, why did we lose it?

It's easy to understand that, during a period when mere survival was next to impossible, making the stones dance would have come to seem unimportant—but not telepathy. We could really have used a power like that during a time of chaos.

People must have felt, as the Younger Dryas erupted around them, that the planet that had always served them was turning against them. The uproar in the sky would have made them certain that the gods they counted on to keep nature in balance had gone to war or gone mad, or both.

The catastrophe would have affected every living soul on the planet. In North America, an entire human community, the Clovis people, was annihilated. Populations plummeted worldwide. And the human mind was changed. As community collapsed, people were forced to focus on their own personal survival as never before. The mastery of great stones no longer mattered. The mastery of the next meal did.

"And the eyes of them both were opened, and they knew that they were naked." That is to say, in their desperation, with each individual struggling to survive, they lost their community of mind. When your next breath may be your last, you focus on your own immediate self. You have to.

Across those awful centuries, they became us. We today are the ghosts of that once-united humanity, and ours is a ghost world.

Before the catastrophe, there is evidence that we were also a telepathic species, linked together in a community of mind that we urgently need to restore.

The idea of a communal human mind isn't just a speculation. There are indications that there was much more relationship between cultures than can be explained by seafaring, even the given the possibility that more sophisticated ships must have existed prior to the Younger Dryas, and probably some distance into the catastrophe as well.

In fact, the Younger Dryas is where communication such as I am discussing here ends. In fact, as it can be seen as early as

Neanderthal times, it would seem to be too ancient to involve long-distance seafaring on a large scale.

The entirely overlooked early example of this sort of communication is the fact that human handprints are found alongside cave paintings across a huge gulf of time and in widely separated parts of the world. These prints were made by blowing a puff of red ochre through a tube against the back of a hand that was pressed against the cave wall. They appear in caves around the world and are present across close to fifty thousand years of human cave painting activity.

I am not arguing that this phenomenon is due to people being able to speak across vast distance with their minds, but rather that there is a telepathic state of mind that we have lost. It was this state that meant, in part, that people far removed from one another would follow the same habits without necessarily being aware of each other. Of course, there were other factors, too, among them the fact that we all share the same brains, therefore the same perceptual systems and the same ideas about reality.

It can also be explained away as a universal practice that arose because it was easy to do and red ocher is plentiful near most of the sites where the handprints appear, and if it was the only example of such a universal practice being repeated among widely separated peoples around the world, that might well be an explanation, an issue to which we shall return.

They have been found in caves in Spain, France, Argentina, Indonesia and Brazil, and it is likely that they appear in others as yet undiscovered. One grouping, in the Cueva de Ardales in Spain, is around 65,000 years old and was probably created by Neanderthals. The most recent grouping so far found is in Cueva de las Manos (Cave of the Hands) in Argentina. There are around

830 of these handprints in that cave, and they are its dominant marking. They were created between 13,000 and 9,000 years ago, during the Younger Dryas.

It is worth asking why there are so many in that one cave, but unfortunately the only answers must be speculative. Most of the hunts depicted in the cave involve guanacos, a form of llama, who were common in the region. At the time, though, the Quaternary extinction was under way in the Americas, and all animal populations were being stressed by violent climate change as the ice age came to an end.

Does the amazing number of handprints, then, represent some sort of ritual effort to attract dwindling or migrating herds—to grasp them, somehow, in the imagination, thus imposing control over them through magical means?

The presence of this specific artifact in so many places over thousands of years suggests not only a universality but also a kind of continuity that we no longer possess in human culture. The most popular argument used to explain the many similar expressions across vast gulfs of time and space is that all human beings share the same brain structures and therefore think the same way. But even the slightest glance at the profound differences between art and artifacts produced in different parts of the world throughout history makes this theory hard to support. The art of ancient Egypt bears little resemblance to that of Sumeria, for example. And yet cave paintings, almost all depicting hunts, are created using very similar techniques worldwide, and bear a striking resemblance to one another. It does seem as if history starts with a divergence from a state of community that had served us for many thousands of years.

After the appearance of historical civilizations, the striking artistic similarity of the Neolithic period ends, save for one notable exception that we will explore shortly.

The art of the Egyptian Old Kingdom and of the Sumerian civilization, which arose at approximately the same time, around 5000 BCE, are radically different. But cave paintings, which were executed worldwide up until the Younger Dryas, and end with the bizarre effort recorded in the Cueva de los Manos, are similar, in some cases to the point of being nearly identical, and yet created across huge stretches of space and time. And, of course, the hand-printing process was the same worldwide.

A case can therefore be made that, as the Younger Dryas came to a close, something changed in the human mind, most particularly in the way widely separated social groups generated similar art.

Cognitive psychologist Stephen Pinker argues that there exist universal cognitive traits that arise out of brain structure and govern such aptitudes as pattern recognition and perception of symmetry, which lead to similarities in artistic expression around the world. But again, the more recent the art, the less similarity there is. At the present time, in painting, for example, individual artists are valued primarily for their personal styles, and artists who adhere to traditions such as realism are considered unimportant. Indeed, the gradual evolution of art throughout history parallels the increasing complexity (and isolation) of the individual. The earliest named artists, such as the Greek sculptor Praxiteles, were renowned for the perfection of the portrayal of the external human form, while a modern portraitist such as Alice Neel or Lucian Freud are valued for their ability to expose the inner life of their subjects, a process that begins in the sixteenth and seventeenth centuries with artists like Leonardo da Vinci and La Gioconda (the Mona Lisa) and Rembrandt, most particularly his later self-portraits.

When we reach back into Neolithic art, we see the human form portrayed, with a few minor exceptions such as the Willendorf Venus, as a sketchy presence. By contrast, the animals who were the objects of the hunt are observed with extraordinary clarity. It can be argued that the human form did not appear because the sense of self—the ego—had not yet evolved. The universal style of the animal portraiture remains strikingly similar worldwide before and during the Younger Dryas. But then it stops. By 5000 BCE, artistic expression is localized. Each individual culture is developing its own art forms, still stylized but no longer nearly as profoundly connected as were artistic expressions during prehistory. The leaping glories of Cretan painting contrast with the ponderous drama of Sumerian art and the elegance of Egyptian expression.

Did some sort of telepathic field exist before the catastrophe, which was among the powers that we lost due to the pressure it put us under?

Carl Jung proposed a collective unconscious that draws on what he called archetypes, which explain the universality of symbols across time and space, but not the universality of animal portraiture that existed in the Stone Age, and certainly nothing remotely like the use of handprints in caves separated by vast distances, as much as 35,000 years in time, and probably even different creator species.

There are a number of siddhis that are relevant to this mystery, as is the modern practice of remote viewing. Having followed remote viewing efforts from various perspectives since its inception,

I am aware of the fragility of the process and the high level of discipline and concentration necessary to actually gain reasonably accurate results, and also that not even the best remote viewers have consistent success. But while it could be related, remote viewing is not an ability of the nonlocal mind that I am suggesting here might once have existed. To me, the evidence is hard to explain by conventional means, although easy to explain away.

The siddhis that are relevant are *Pratibha*, which opens the door to higher knowledge; *Dura Drishti*, which is distant vision, akin to remote viewing; and *Divya Drishti*, which is clairvoyance. But none of these are permanent states. They are, rather, conditional on a high level of spiritual attainment and extreme concentration. This spiritual attainment, universally, involves the containment of ego and the transfer of the attention to a region of consciousness that lies outside of the boundaries of self.

It would seem necessary for the mind to have been in another state for this to have happened as it did, and there is one theory that might speak to this possibility. It appears in Julian Jaynes's 1976 book *The Origin of Consciousness in the Breakdown of the Bicameral Mind*. Jaynes asserts that the human mind was different prior to the beginning of history, and into first few hundred years of the historical era. He theorizes that the two hemispheres of the brain communicated with one another in the distant past by means of auditory hallucinations which were transmitted from one side of the brain to the other and perceived by individuals as the voices of gods. Only as more complex societies developed did people begin to need to be separate individuals. When populations rose and cities developed, challenges like protecting hard-won possessions and securing one's place in the hierarchy of society required a sense of personal boundaries and self-awareness—that is to say, an ego. The authority in the head fell silent, giving way to this other manager of needs and rights. In the appearance of the first cities, the first culturally unique art, laws, language and writing, what we are seeing is the beginning of the individual.

Jaynes makes a carefully considered case for his ideas, but he received a significant amount of criticism because of a lack of hard evidence. What form that might have taken, though, beyond what he presents, escapes me. While his case for the precise way that the mind changed may or may not have substance, the fact, as he points out, that the mind did change is hard to dispute, just looking at the way civilization suddenly appeared in numerous different places around the world at roughly the same time.

But what of those civilizing gods? Who were they and where did they come from? Surely, if they didn't exist at all, they wouldn't feature in so many accounts of the founding of human societies.

Many cultures have origin stories about either gods or wise strangers who brought them what is generally described as 'civilization.' This has two aspects. First, they are bringers of techniques, such as farming, which was brought to the Egyptians by Isis during a time when they were starving and had turned to cannibalism. This is probably a memory of the first years of the Younger Dryas, when dramatic climate disruptions interrupted age-old hunting practices and a long period of protein deprivation ensued.

As the glaciers melted, the Akhet, the Nile flood, became regular, and the Isis story likely emerged as a cultural artifact to explain a time when the women, who might have been experiencing the use of their children as a food source—and themselves when they ceased to be able to bear—developed seasonal agriculture to provide a more acceptable one.

Similarly, the Sumerians were apparently wandering hunter-gatherers prior to the appearance of their first urban communities in the fertile crescent between the Tigris and Euphrates rivers. Like the Egyptians, the disappearance of game forced them into larger communities that could support labor-intensive agriculture. They personalized this process by identifying it with strangers called Apkallu, which were variously described as sages or gods who wore the skins of fish. Perhaps what this really records is an encounter with people who were forced to come up from flooding communities at lower sea levels, literally coming up out of the water as far as these highland wanderers were concerned. They understood them as being related to the creatures of the sea from which they had emerged, bringing with them knowledge of the higher-level social organizations in which they had lived, and the techniques needed to rebuild them. They saved the hunter-gatherers from a catastrophic game scarcity by turning them into farmers.

In Mesoamerica and Peru, when the Aztec and Incan empires were invaded by the Spaniards, an effort was made to record their myths, beliefs and histories even as the conquerors brutalized them and destroyed their civilizations. Both the Incas and the Aztecs appear to have had stories of gods who brought civilization to them. In the case of the Inca, their main god, Viracocha, was said to be bearded and white and, like the Apkallu, to have come from the sea. (But not, fortunately for him, wearing a fish skin.) Likewise, the important Mesoamerican deity, Quetzalcoatl, is portrayed in some cases as a white, bearded figure who came out of the sea, bringing civilizing notions with him. Because Cortez was bearded and white and came from the sea during a "One-Reed Year," which was a period in which the return of the god was believed to be possible, the Aztecs were at first confused about the identity of the conquistador. However, the surviving Indigenous people, who, a hundred years later, numbered just 1 to 3 million, down from 15–30 million prior to the conquest, were unlikely to be under any illusions.

One is tempted to consider that these benevolent white sages were conflated with the Spaniards, but there is pre-conquest statuary of both Quetzalcoatl and Viracocha in their human forms that show them as being bearded and tall. As to white, that's an open question. All references to their color come later.

Were they survivors of a civilization that was drowned during the Younger Dryas, leaving behind the Atlantis legends? Perhaps, but the Apkallu appear to predate Quetzalcoatl and the Viracocha by thousands of years. Or do they? The first images of the feathered serpent appear at Teotihuacan near Mexico City. An alternative archaeologist, Dr. Mark C. Carlotto, argues in his book *Before Atlantis* that Teotihuacan would have been aligned to true north if the pole had once been located where Hudson Bay is now. If that is the case, the site must be much older than is currently supposed. The notion of crustal shift is controversial, but the concept of true polar wander isn't in dispute. However, it is thought that the slow wander of the poles takes place over millions of years. During the collapse of the ice sheet, when the planet's surface mass was shifting dramatically away from the poles due to rapid ice melt, there could have been a sudden wander, which was misunderstood by Charles Hapgood, the originator of Earth Crustal Displacement Theory, as something that was due to unknown geologic forces when it actually takes place during the huge shifts in mass that occur at the end of ice ages. Given that our planet has experienced something like twenty ice ages during the last 3 million years, if the mass shifts are causing rapid wanders, they would be a frequent occurrence. That this may be so is suggested by the fact that ancient sites are clustered in terms of their orientations. Groups of them are oriented, for no apparent reason, toward different northern positions that are not poles. It is hard to not to conclude that the reason is that the poles were once in those locations. I might add that it is not magnetic north, but true north, physical north, that is involved.

For example, there are no fewer than 29 sites worldwide, Teotihuacan included, that are all oriented toward what is now Hudson Bay. One of the very oldest of all structures on the planet, the Osirion, is oriented toward Hudson Bay, according to Carlotto. Conventionally, the Osirion is dated to the reign of the pharaoh Seti 1 (1290–1279 BCE), but the monoliths out of which it is constructed are far larger than the stones used in the nearby Temple of Seti I, which is the basis for its dating.

Does this mean, then, that sudden polar wanders do occur? There are so many other examples of groups of ancient structures being oriented to northern points that are no longer the North Pole that it would seem that sudden wander might be the rule rather than the exception.

If this is correct, then Teotihuacan must be very old, probably as old as the Osirion, and, very likely, other megalithic sites around the world. These structures would, like Baalbek, all have been built at a time when we could move massive stones in the same way the Ed Leedskalnin did the smaller ones that make up his construction. In a time, in other words, before the struggle of the Younger Dryas caused the human mind to fall into the crippled state from which it has yet to recover.

If we hypothesize that all of the sites in the world containing gigantic blocks that we would be hard put to move today belonged to the same proto-civilization, then it was clearly world-girdling, very ancient, astonishingly capable and probably linked by the same community of mind that inspired the handprints.

When I say "astonishingly capable," I refer, of course, to the astonishing artifacts that it has left behind, from Gunung Padang in Indonesia to Göbekli Tepe, Nan Madol, Baalbek, the Osirion and many other massive sites. Probably all the structures on the Giza Plateau are involved as well, and sites in the Americas such as Sacsayhuaman in Peru, and again, many others.

Somebody must have built the many structures around the world that we cannot explain and could not easily build now. Obviously, given that similar techniques, all unknown, were used worldwide, this must have been some sort of civilization, but it is not my purpose here to advocate for it. As far as I'm concerned, the facts render that argument meaningless. It was clearly there and now it's not. During the Younger Dryas, there were vast upheavals, fires and floods, some of them huge and sudden. It must have been destroyed by this. But I don't think that we are necessarily going to find cities such as we think of them. Whatever was going on in that civilization, it required the creation of massive monuments that now seem to have no clear purpose. So there was something about it that we do not understand, that is based on a different way of thinking from our own. To even begin to understand how it conceived of itself, we should return to the very earliest stories. Perhaps we will find some hints there, and, if we are open-minded and careful, be able to use those hints to recover the long-lost ways of thinking that caused the platform at Baalbek and the Osirion and the other mysterious monuments to be constructed. Whatever the reason, the sheer magnitude of these undertakings means that they were extremely important to their creators.

I am much more interested in the discussion of the powers the civilization possessed, why they have been lost, and how they can be recovered.

Not only did they apparently possess what to us seem to be magical powers, judging from what remains in Egypt and many of the other sites, they also had brilliant tools, such as exquisitely hard drill bits and powerful rock saws, and the means to deliver energy to them. They did not have the sorts of engines we would use, or we would have found their remains.

They were master engineers and craftsmen—in fact, their mastery was greater than our own in the areas of stonecutting, the movement and fitting of extremely heavy stones, and the carving of smaller objects to extremely precise tolerances. And yet, they left no tools. Given the hardness of the drill bits and saw blades they must have used, this is difficult to explain. Diamond drill bits are not going to erode, and the dry sands of Egypt preserve even papyrus, so where are they? There are only three possible answers, as far as I can see: drowned beneath the floods; hidden in the desert; taken off the planet. Unfortunately, as we know so little about what was actually happening then, there is no way to speculate usefully any further.

That there was a crisis, and that its memory still lingered in the fifth century BCE when Plato was writing, seems to me now to be beyond dispute. The crisis is accurately described and dated in Plato's dialogues the Critias and the Timaeus, written at that time. These have been assumed to be mythological stories, but now that we know more about the dating of the Younger Dryas and how it unfolded, it is clear that Plato was reporting something that had been handed down in memory and not confabulated.

In the Timaeus, it is said that a force from somewhere beyond the Mediterranean attempted to conquer the Mediterranean region and was rebuffed by Athens. A "great conflagration" is mentioned, as well as a land mass that sank abruptly into the ocean. This is all dated to 9,000 years prior to the writing of the dialogues, which would be about the time of the inception of the Younger Dryas. The event started with the great North American fire, then the sudden collapse of the Laurentide glacier. Fire, then flood. The glacier was struck by something so powerful that it sent backsplash from the region that is now the Great Lakes into the Carolinas, and caused gigantic clumps of ice to sail across the continent into New Mexico.

If a more developed civilization existed near the oceans then, it would have experienced some very sudden inundations as a result of this event. It wasn't, as Plato assumed, that it sank into the sea, but rather that the sea rose and submerged it.

It might well have made a desperate attempt to save itself by invading the Mediterranean. That it was defeated by a far less advanced group in the form of fighters from neolithic Athens makes it seem as if it must have been a remnant force, and probably also come from a community that was not practiced in warfare. That whoever sent it knew the region is likely, as Egypt and the Levant, and to a lesser extent Türkiye, are, as we have seen, covered with mysterious remains.

One of the sites, which was constructed during the unfolding of the Younger Dryas, Göbekli Tepe, reveals something that is, in one way, very much like the ubiquitous handprints. This form appears at the top of Pillar 43, the Vulture Stone, and consists of three square objects with arches above them. What is so extraordinary about this carving is that the same form appears in so many different places and times. There is no way to determine whether or not these depictions had the same significance in the different societies that created them, but the use of this particular form suggests some sort of connection. The oldest of the images may be at Göbekli Tepe, but some are also seen in rock art in the Americas and Australia. They appear as well in the various Middle Eastern cultures, including Sumeria, Assyria, the Hittite culture, Babylon and Persia. They also appear in Olmec reliefs from as recent as 400 BCE, and in other Mesoamerican cultures from even more recent periods. This means that they are present in many cultures across the world for 8,000 to 9,000 years.

As I have said, there is no way to determine their purpose or meaning, or whether or not they had some sort of actual power beyond their symbolic or talismanic significance. It can only be said with certainty that they are associated in many cultures with the appearance of early creative or inciting presences such as the Apkallu in the Middle East; in Indonesia on the Island of Sumba, the progenitor sage Panji; and the civilizing gods of Mesoamerica. In addition to their identification with culture building, they are associated with fertility, as is demonstrated by their connection to the rain god Tlaloc in Mesoamerica and the Tree of Life in Assyria.

Were all of these cultures, then, connected, from Göbekli Tepe to Sumba to Sumer and Assyria, and many more, or is the explanation closer to what I have proposed to explain the universality of cave art handprints, that there was in the past a community of mind that was shattered by the chaos of the Younger Dryas and the disruptions and population declines that came with it?

I have not yet specifically discussed healing, which is what the Varginha alien did to demonstrate to us what we have lost. Early medical texts from the Edwin Smith Papyrus (Egypt, circa 1600 BCE) on through the classical period often mix surgery, healing concoctions and magic, but they do not illustrate the Varginha alien's ability to self-heal. As I pointed out in *Jesus: A New Vision*, the practical magic of the ancient world was centered around the placebo effect. Because it was not understood in those days, it could be powerful medicine. But it was nothing like what the Varginha alien did, and there is no evidence remaining that we were ever proficient at such dramatic and effective magical healing as he demonstrated.

But could we, perhaps, gain this power?

That question is unresolved, but if we are to save ourselves and our planet from the chaos that is racing toward us right now out of the darkening skies of the future, we should try to answer it, and as soon as we can.

CHAPTER SEVENTEEN

OUR WORLD TODAY—AND TOMORROW

We are in the early stages of a crisis right now that is in important ways as dangerous as that of the Younger Dryas, and our survival is as much in question now as it was then—at least as much. So far, it seems that it will unfold with less drama than that crisis, but its effects are going to be just as profound. Some sort of civilization was destroyed by the Younger Dryas. It was not like ours, I wouldn't think, with great cities that stood in defiance of nature. Rather, to build its great structures, it used powers of mind and Earth energies where we use construction equipment and heat. Its technology was mind based. Ours is fire based. I think that we are this way because it was destroyed very suddenly by the fearsome floods that followed the collapse of the Laurentide glacier, which was caused by a series of impacts and probably took place over a matter of weeks or months. This came after years of slower flooding, and followed weather-disrupting smoke pollution from the immense North American fires.

The civilization, never large by our standards, would have been seriously weakened by the collapse of food sources due to the fact that the smoke pollution would have done exactly what it did in 536 CE, which was to shut down the growth cycle of plants due to lack of sunlight and warmth. The population, weakened by malnutrition, would have then become prey to disease. Then this struggling population, among whom would have been concentrated most of the people who could use their minds in the ways I have been discussing, would have been overwhelmed by enormous, planet-wide tsunamis. These populations, concentrated along the then-existing coasts, would have been drowned.

We remember all of this in our fire and flood myths. In them, we anthromorphized natural phenomena first into battles among sky-dwelling deities, then later into punishment of mankind by flood.

The truth, I think, is the one thing that we find hardest to bear. No gods or aliens were involved. We fell victim to a natural catastrophe, which was governed by the real deity that rules over this universe, which is chance.

But that isn't all there is to it. There is also the possibility that a hidden mind is involved, and has been for a very, very long time. I say possibility because the first part of what I am about to describe could be a matter of chance, as improbable as that may seem. But this is a big universe and, from our own experience, it seems clear that chance is a major driving force.

Our solar system can be seen in two ways. The first is that it is one of many trillions, and fell together as it did purely on the basis of chance. The second is that it has been designed to draw life out of one of its rocky inner planets, which would be ours.

But why use that word "designed"? Is there any reason to believe this? Actually, there is. The anthropic principle states that the structure of the solar system is so perfectly designed for life that it cannot have been accidental.

Earth is within the 75-million-mile-wide habitable zone. It has, uniquely for this solar system, a moon large enough to govern its tides and reduce the velocity of its rotational winds to a speed that enables complex life forms to develop. Its moon is also precisely the right size to perfectly shadow the sun during a total eclipse. The gas giants in the outer reaches of the solar system attract a great deal of debris into the powerful gravity wells, and the Moon removes more of the debris that reaches the vicinity of Earth. While this doesn't protect the planet completely from impacts, it reduces them dramatically. The oceans of Earth, which actually form a thin film on the surface, should have evaporated long ago, but they show no sign of doing this.

When taken together, it is hard to believe that some sort of design wasn't involved. There is another factor, not much mentioned in the various designer arguments that may support the idea. This concerns one of seven of the most important physical constants, known as the fine structure constant. It is 1/137th. But why that particular number? There is no known reason for it. It describes the strength of the electromagnetic force, which governs how light interacts with particles. It is on this constant that all structure depends. Unlike the other constants, though, there is no clear reason that it is this particular number, which troubled the twentieth-century physicist Wolfgang Pauli so much that he initiated a correspondence with the mystical psychologist Carl Jung in order to attempt to find something, perhaps in the mysteries of the spirit, that would explain it.

It has also been argued that if the value of the fine structure constant was any different, there could be no life in the universe, or that life would be very different and probably not as complex as what we find all around us. While this is debatable, there is no question that the particular value of the constant governs the structure of life and everything in the universe.

If the value was higher, for example, the strength of the bonds between electrons and the nuclei of atoms would be greater, meaning that chemicals essential to life could not form. If it was lower, matter would lack fundamental stability. Just as the careful values of size and distance make the solar system seem designed, the exact value of the fine structure constant seems intentionally chosen.

It can therefore be realistically argued that there must be some sort of design inherent to the universe. Further, as Max Tegmark shows in his book *Our Mathematical Universe*, the formulae that govern the structure of reality had to have existed before the emergence of the universe. That is to say, the math that governs the way the universe works, how it forms stars, how they evolve—how it all works—must have existed before it could have.

So where was all this math? In the mind of God, perhaps? But what is that? Is it conscious in some way that we can relate to, or so distant from us that it will never be more than a question?

Does God, in short, have anything to do with us and what we are?

Oddly enough, there is a number that suggests that this solar system was indeed designed, and not only that, that somebody in the distant past understood this and left a way for us to hold on to the memory.

The diameter of the Earth times 108 equals the diameter of the Sun. The diameter of the Sun times 108 equals the distance between the Sun and the Earth. The diameter of the Moon times 108 equals the distance between the Earth and the Moon. The number of beads on a Hindu prayer garland, or mala, is 108. The earliest mention of a string of beads being used in prayer appears in the Rig Veda, circa 800 BCE, but the actual origin of the garland and the reason that it contains 108 beads is unknown. But given that this number also reveals the design of the relationship between Sun, Earth and Moon, it's not unreasonable that it was intended in the distant past to be a mnemonic device.

Looking at the history of Earth, if an über-consciousness exists, it cannot be said to be benign, not given our planet's violent history and the peril that we now face. After all, it seems to have allowed, or even possibly instigated, the destruction of a way of human life that seems to have been in advance of what we have now. If one takes the old texts at face value, this happened because of what amounts to the creator's jealousy. In fact, the stories of the many manifestations of God read like a soap opera underscored by a particularly violent thriller—but that is only if you confine yourself to one perspective. There is always another, the classic example being the destruction of the dinosaurs.

They would not think of the Chicxulub asteroid as in any way a good thing. But our attitude is different. Were it not for the shock of that impact, we probably wouldn't exist, so for us it was nothing less than a miracle. Extinction events inevitably lead to evolutionary innovation. Challenges result in growth in human life as well. We entered the last ice age living in trees and came out of it organized into tribal units, covered with animal furs, bearing effective weapons and accompanied by dogs who aided us in hunting. Had there been no deep freeze, we would probably have remained in the trees.

The random violence of the universe enforces change, and is thus also the engine of evolution.

But is this all inevitable? Evolutionary innovation can be explained. It's not a miracle, but an inevitable result of the fact that the most adaptable species are going to be the ones who survive an extinction-level event. They will inevitably fill the life niches left by the event's victims.

The first ice age, the one that began the Pleistocene, was preceded by a period known as the Pliocene. This period lasted approximately 2.7 million years. Earth's climate was warmer and more stable than it is now, and life in that mild climate does not show much evolutionary change. But then the ice age cycles of warmth and cold start, and along with them comes a virtual riot of evolutionary change as creatures seek to adapt to the roller-coaster climate. In general, though, the total number of species has been in decline throughout the whole 2.8 million year period, and we are seeing the climax of that decline right now.

In fact, we are in the climax of an extinction event that began not with the Younger Dryas but rather with the start of the cycle of ice ages itself. Somebody, though, seems to have understood how the periods of cold followed by brief warm periods (interglacials) work, and how the interglacial we are in now, in particular, will unfold.

The Western astrological calendar dates from the Babylonian period around the second millennium BCE, and the structure we use today, from Egypt during the Hellenistic period, which began in the fourth century BCE.

This calendar is divided into 12 periods, each of which covers about 2,000 years and is keyed to the precession of the equinoxes, which is a slow shift in Earth's rotational axis. This shift causes the north star to change as the pole circles a central point. The 12 houses of the astrological calendar mark when, in the long count of the precession, each constellation will be ascendant, and the animals associated with them also relate to conditions during that particular 2,000-year age.

At the present time, we are moving from Pisces to Aquarius. While we were in Pisces, we were a fish swimming in the waters of the Earth, or, to put it another way, an embryo living in Earth's womb and being provided by her with everything we needed. Now however, Aquarius is pouring out the waters and the little fish is about to end up on dry land.

That is to say, Earth's waters have broken and we're headed down the birth canal. We are either going to be born dead or alive, and one wonders, and with more than a little urgency, which it will be.

Except fish have already done it. Some 350 million years ago, some of them somehow managed to survive in shallow tidal pools for a sufficient number of generations to evolve nostrils along the top of their flat heads, adding them to their ancient gills and enabling them to live through periods when the water level was low. Later, when the Permian extinction took place, with their greater adaptability, they were among the survivors.

I want us to become like these amphibians—not to grow gills, but to become cleverer than we are.

There are two ways of doing this, and, given how fast things are changing, I think that both are probably necessary if we are going to avoid a great decline of our species, possibly even an extinction. As to the role our visitors might be playing in this drama, putting aside ideas of their being demons or some kind of interstellar invasion force, their real purpose here is quite clear: they propose to be midwives to our birth—and I, for one, propose to help them do that.

It won't be easy, and this small but steely tough band of midwives who are here are not going to give us more than the absolute minimum amount of help necessary. If we fail, then we fail.

There are two ways of insuring our future, one minor and one major. The minor one is the only one we actually think about. It involves cleaning up our air and our environment and creating more, and more powerful, technologies in support of that effort. The major one is, of course, re-igniting our lost powers. The former we might be able to do; the latter seems impossible.

But is it?

However it happened, for the most part we lost those powers during the Younger Dryas. Probably most of the people who could use them died when the lowlands of the world were abruptly flooded during the sudden collapse of the glaciers. Given the chaos of the period, though, there is no certain way to know. Except that it certainly happened. The conventional explanations for the megalithic structures all around us simply do not work. In this case, alternative archaeology seems to have the more probable theory. A powerful civilization was destroyed by the violence of the Younger Dryas.

The word 'civilization' conjures images of great, glittering cities, but I don't think that this is what was there. The way the ruins that we do see are built points not to great engines capable of moving hundred-ton stone monoliths or transporting millions of heavy basalt logs, but to a much simpler and more elegant process.

It seems possible that G. I. Gurdjieff's idea, that some sort of connection between us and earth energies was severed, may be closer to the truth. What was it, though? I think that our bodies and our planet are probably willing to tell us if we ask the right questions.

First, I think we have to get past the idea that acquiring the power of the siddhis would take a special sort of person. On the contrary, I think that the exact opposite is true. It is going to be the average person who can do this. Not being able to will be what is unusual.

A 2022 study done by Helané Wahbeh, Dean Radin and colleagues called "Genetics of Psychic Ability" supports this as a real, if unexpected, possibility. The study involved carrying out DNA analysis of a group of proven psychics and comparing the results to the DNA of a control group who had no discernible psychic ability. The test of psychic ability was as rigorous as such a test can be.

A surprising difference was observed between the psychics and the people without psychic ability. Almost all of the non-psychic controls contained a variant—technically, a single-nucleotide polymorphism (SNP)—that is absent in most people, including all the psychics in the study. In other words, the study—admittedly preliminary—suggests that psychic ability is the normal human condition, and psychic silence is not (Wahbeh et al. 2022). It is, perhaps, a mutation. This means, of course, that one of the first things we need to do is to stop assuming that we are less than what we are. The catastrophe we went through, which overwhelmed all of our powers with its fury, didn't strip them away. They are still there, present in our genes. What we lost was our confidence, and that is what we must regain.

The study also shows that certain religions select against psychic ability, most notably Christianity, which spent hundreds of years persecuting people with psychic abilities, calling them witches, pagans and so forth, and killing them for being in league with demons. The result of this was that, in areas where these activities were aggressively pursued for long periods of time, the commonplace psychic variant became rarer, and the populations ceased to have this normally ordinary ability. Central Europe, where the Holy Roman Empire governed for hundreds of years and aggressively pursued the elimination of people with psychic powers, now has a higher proportion of people with the non-psychic variant. In other words, this religious group blinded itself, and handed its deformity down the generations.

It is one thing to possess psychic power as a potential, though, and another to actually use it. Judging from the stories of the levitators, it seems clear that the power is much greater in people who are in an ecstatic state. Such people can change their physical status profoundly, transforming a heavy body, complete with its clothing, to something that is lighter than air.

As our scientific culture denies absolutely that such a thing is possible, we have never attempted even to theorize about why it might happen. As we no longer have access to people who can readily levitate, we have no subjects to test. All we can do right now is to model what has to be true of a person's physical body for it to happen. Once we have determined that, then perhaps we can make some progress inducing it in test subjects. If that ever happens and becomes well-known to the public, perhaps the ability will spread.

This goes for the more important siddhis as well. We have first to embrace the idea that they are real, and it is possible for us to learn to use them, before we can have any chance at all to master them and use them in the world.

Our visitors have shown us that they are real. Many people have experienced telepathy with them. Some have experienced levitation while close to them, as if proximity can spread it. A few have experienced healing.

Could we become a community of minds?

Could we learn to control the density of our bodies?

Could we connect so deeply with ourselves and our planet that we would become able to heal her and heal ourselves?

Could we project our consciousness, even our physical bodies, into other worlds and other realms of reality? Or, put another way, is it our destiny to successfully be born of Earth and into the great cosmos, or are we doomed to die here, an infant not quite strong enough to survive?

It is early evening in the quiet little town where I am writing this. The stars are just coming out. Far to the north, an elegant plume of smoke warns of another fire. Restless weather troubles the Gulf of Mexico; a drought I predicted in my 1987 book *Nature's End* threatens to take hold in the Midwest.

I am just recently back from India, which I found to be the poorest, most crowded and spiritually vivid place I have ever been. My chief memory of that place is the most human thing about it, which is that it cannot possibly work—and yet it does.

We cannot possibly do what we need to do, and meet the challenge laid before us by the Varginha alien. Or is that wrong? If our visitors didn't think that we could be born successfully, I doubt they'd spend two minutes on us.

I pause in my work, lean back from my desk and close my eyes. I wonder 'what is next, here?' Then, quite suddenly, I know: My words have come to their whispered end.

They flow into the leaves in the trees outside my window, joining themselves to the sigh of the night wind. I become aware, also, of the excited cries of the neighbor children playing games in the shadow-filled walkway. Will they one day open their thoughts to one another so completely that they form a new kind of human community? Or is it more a matter of noticing that this is already true, and accepting it? Will those miraculous young people join their new community to others in the stars, and enter, at last, into communion with worlds beyond?

Maybe this book is my last throw or maybe there's to be more. I don't know, but I do know this: I have given everything I now have to those children, but if I can find more, I will give more.

A woman across the way is singing. A dog barks with excitement. I know that dog, his name is Billy. He belongs to the children and he is, as always, joyous.

The moon sails the clouds. My heart is touched by its flight and the wind's words and the singing and the barking and the children's excited cries. I know this feeling. It is the grace of ordinary life and what I feel in it, and in all the world, is the mystery, the wonder.

The End

APPENDIX 1

2002 SEALED AFFIDAVIT OF
WALTER G. HAUT

DATE: December 26, 2002
WITNESS: Chris Xxxxx
NOTARY: Beverlee Morgan
U have,
My name is Walter G. Haut.
I was born on June 2, 1922.
My address is 1405 W. 7th Street, Roswell, NM 88203.
I am retired.

In July, 1947, I was stationed at the Roswell Army Air Base in Roswell, New Mexico, serving as the base Public Information Officer. I had spent the 4th of July weekend (Saturday, the 5th, and Sunday, the 6th) at my private residence about 10 miles north of the base, which was located south of town.

I was aware that someone had reported the remains of a downed vehicle by midmorning after my return to duty at the base on Monday, July 7. I was aware that Major Jesse A. Marcel, head of intelligence, was sent by the base commander, Col. William Blanchard, to investigate.

By late in the afternoon that same day, I would learn that additional civilian reports came in regarding a second site just north of Roswell. I would spend the better part of the day attending to my regular duties hearing little if anything more.

On Tuesday morning, July 8, I would attend the regularly scheduled staff meeting at 7:30 a.m. Besides Blanchard, Marcel; CID Capt. Sheridan Cavitt; Col. James I. Hopkins, the operations officer; Major Patrick Saunders, the base adjutant; Major Isadore Brown the personnel officer; Lt. Col. Ulysses S. Nero, the supply officer; and from Carswell AAF in Fort Worth, Texas, Blanchard's boss, Brig. Gen. Roger Ramey and his chief of staff, Col. Thomas J. DuBose were also in attendance. The main topic of discussion was reported by Marcel and Cavitt regarding an extensive debris field in Lincoln County approx. 75 miles NW of Roswell. A preliminary briefing was provided by Blanchard about the second site approx. 40 miles north of town. Samples of wreckage were passed around the table. It was unlike any material I had or have ever seen in my life. Pieces, which resembled metal foil, paper thin yet extremely strong, and pieces with unusual markings along their length were handed from man to man, each voicing their opinion. No one was able to identify the crash debris.

One of the main concerns discussed at the meeting was whether we should go public or not with the discovery. Gen. Ramey proposed a plan, which I believe originated with his bosses at the Pentagon. Attention needed to be diverted from the more important site north of town by acknowledging the other location. Too many civilians were already involved and the press already was informed. I was not completely informed how this would be accomplished.

At approximately 9:30 a.m. Col. Blanchard phoned my office and dictated the press release of having in our possession a flying disc, coming from a ranch northwest of Roswell and Marcel flying the material to higher headquarters. I was to deliver the news release to radio stations KGFL and KSWS, and newspapers the Daily Record and the Morning Dispatch.

By the time the news had hit the wire services, my office was inundated with phone calls from around the world. Messages stacked up on my desk, and rather than deal with the media concern, Col. Blanchard suggested that I go home and "hide out."

Before leaving the base, Col. Blanchard took me personally to Building 84, a B-29 hangar located on the east side of the tarmac. Upon first approaching the building, I observed that it was under heavy guard both outside and inside. Once inside, I was permitted from a safe distance to first observe the object jus recovered north of town. It was approx. 12-15 feet in length, not quite as wide, about 6 feet high and more of an egg shape. Lighting was poor, but its surface did appear metallic. No windows, portholes, wings, tail section, or landing gear were visible.

Also from a distance, I was able to see a couple of bodies under a canvas tarpaulin. Only the heads extended beyond the covering, and I was not able to make out any features. The heads did appear larger than normal and the contour of the canvas over the bodies suggested the size of a 10 year old child. At a later date in Blanchard's office, he would extend his arm about 4 feet above the floor to indicate the height.

I was informed of a temporary morgue set up to accommodate the recovered bodies.

I was informed that the wreckage was not "hot" (radioactive).

Upon his return from Fort Worth, Major Marcel described to me taking pieces of the wreckage to Gen. Ramey's office and after returning from a map room, finding the remains of a weather balloon and a radar kite substituted while he was out of the room. Marcel was very upset over the situation. We would not discuss it again.

I would be allowed to make at least one visit to one of the recovery sites during the military cleanup. I would return to the base with some of the wreckage which I would display in my office.

I was aware two separate teams would return to each site months later for periodic searches for any remaining evidence.

I am convinced that what I personally observed was some type of craft and its crew from outer space.

I have not been paid or given anything of value to make this statement, and it is the truth to the best of my recollection.

THIS STATEMENT IS TO REMAIN SEALED AND SECURED UNTIL THE TIME OF MY DEATH, AT WHICH TIME MY SURVIVING FAMILY WILL DETERMINE ITS DISPOSITION.

Signed: Walter G. Haut
Signature Witnessed by: Chris Xxxxxx
Dated: December 26, 2002
Source: Tom Carey & Donald Schmit, Witness to Roswell, 2007

APPENDIX 2

The Reddit Document

From the late 2000s to the mid-2010s, I worked as a molecular biologist for a national security contractor in a program to study Exo-Biospheric-Organisms (EBO). I will share with you a lot of information on this subject. Feel free to ask questions or ask for clarification.

It seems like all my comments are being deleted. I will post answer at the end of the message.

From the late 2000s to the mid-2010s, I worked as a molecular biologist for a national security contractor in a program to study Exo-Biospheric-Organisms (EBO). The aim of the program was to elucidate the genome and proteome basis of these organisms. Although the study of OBCs has been going on for decades in other programs, the new high-throughput DNA sequencing technologies of the late 90s unblocked stagnant research in this area. Since then, several breakthroughs have led to significant advances in our understanding of the genome and proteome of these beings. What we've learned so far has enabled us to outline some disconcerting perspectives about our place in this universe. Briefly, we've discovered that the EBO genome is a chimera of genomes from our biosphere and from an unknown one. They are artificial, ephemeral and disposable organisms created for a purpose that still partially eludes us. I'll be substantiating my statements after a brief introduction.

The reason for disclosing these secrets is quite simple. I believe that every human being has the right to know the truth, and that to progress, humanity needs to divest itself of certain institutions and organizations that will probably not survive these revelations in the long term. I'm aware that I'll have very little impact in this regard, but I still believe that small leaks are necessary to break the dam of misinformation on this subject. When the governments will eventually reveal these secrets, there will undoubtedly be a societal upheaval, but in my opinion, the longer we wait, the worse it will be. I choose to divulge what I know anonymously out of selfishness for the well-being of myself and my family. I'm aware that this diminishes the reach and credibility of my message, but it's the furthest I am willing to go. I chose this forum because it offers a good compromise between anonymity and popularity. In order to protect my anonymity, I will be purposely vague or even contradictory about any information that could identify me (date, education, role etc.). I'll even introduce red herrings in this respect. I want to make it clear that any information related to the subject of the research will not be treated in this way.

Before going any further, please excuse me if you find it difficult to understand what I'm explaining. Some parts of my text are very technical. It's difficult to find the right balance between vulgarization and scientific explanation. I'll continue by talking about myself. What's the point of talking about me knowing that the information will necessarily be misleading? I simply want to introduce a perspective on the type of people who work there, normal scientists. I have a Ph.D. in molecular biology. I didn't actively seek to be part of this program, rather it was a stroke of luck that introduced me to one of the senior scientists. I met this person at a conference where I was presenting a poster on my Ph.D. research. When I think back, I don't believe he was impressed by what I was presenting, because it was quite frankly a project that wasn't going anywhere. I think it was rather the most important aspect of a professional life: the attitude and the ease with which you make connections. Shortly afterwards, I graduated and received a call from this person offering me a position. At the time, everything pointed to me working in a regular laboratory.

I did a series of three increasingly suspicious interviews, each in a different location, where my scientific background and knowledge became less and less relevant. The first was with two of the senior scientists, the second and third with people I've never seen again and who were obviously not interested in science. Sometime after the interview, I was asked to go to a fourth location where what seemed like a corporate lawyer presented me with an NDA. He made sure not only to explain every detail, but also that I understood the consequence of not respecting it.

The first Employment weeks were by far the most memorable, although I spent most of that time in a depressing archive room. It consists almost exclusively of reading about the subject of study and to get us up to speed. There's no secret Wikipedia or even a reference book to guide us. There are only dry reports, memos, presentations, procedures and SOPs. These documents are almost exclusively about the biology of EBOs, but there are also a few that deal with other subjects such as their food, religion or culture. There were no documents on their technology.

As mentioned above, the aim of the project is to gain a better understanding of the EBO genome and proteome. To achieve this, a team of around twenty scientists, four senior scientists and a director was involved. The scientists, like myself, had as their main responsibility to carry out the technical work. As each scientist had to my knowledge a Ph.D., we were all somewhat overqualified for what is ultimately a technician's job. The senior scientists, who make full use of their diplomas, had the task of designing the assays and had a supervisory responsibility. They were also in charge of training new employees, and sometimes even came in to do technical work. The director, of course, was the person in charge who dictated priorities to the senior scientists. He was rarely on site, and the few times he was, it was to attend meetings. Other than the scientific staff, there were security guards working for one subcontractor or another. There were no support staff such as janitors or maintenance workers. Scientists were responsible for this kind of work. In addition, logistical constraints ensure that every scientist is capable of carrying out any technical activity.

The laboratory itself is located in Fort Detrick, Maryland, in a building used for legitimate biomedical research. The clandestine operations are carried out in a restricted part of the basement, out of sight from regular workers. Contrary to what one might imagine, the biosafety level is not maximal for this type of research. Indeed, the lab containing EBO samples or derived cell cultures is BSL3, while the lab where assays are conducted are only BSL2. The BSL3 area of the facility includes a freezer room and a cell culture lab and is only accessible through an antechamber from the BSL2 section. EBO carcasses are preserved in horizontal freezers at a temperature of -80°C nominal. To maximize the preservation of these carcasses, they are preserved in vacuum bags and the air in the room is controlled to minimize humidity. There are only four bodies and none of them are complete. It's obvious that these creatures have died as a result of major trauma. I've never witnessed a motorcycle accident fatality, but it probably looks similar to this. It is acknowledged that there are more EBOs carcasses at other locations. The cell culture laboratory, as its name suggests, is where cell lines derived from EBOs are grown and related activities are performed. I'll talk in more detail about these specific cell lines later on. The BSL2 part is mainly used for assays, immunohistochemistry, genetic engineering, immunocytochemistry, storage etc. There's also a cell culture lab, but this is used for more traditional cell lines. Other than the labs, there are all the amenities you could find in an office. Note that the internet access is limited to senior staff and up. There is, however, an intranet for bioinformatics needs.

On the subject of the biology of these beings, I'll start by discussing genetics, then their gross anatomy and finally their biological systems. For the sake of clarity, the information that I provide here is an aggregation of what I have observed and what I have read. I will make many comparisons with human anatomy because it is the most logical reference.

Genetics:

First, I'd like to discuss their genetics. Their genetics are like ours, based on DNA. This fact was very puzzling for me when I first learned about it. We imagine that beings from an alternate biosphere would have genetics based on a completely foreign biochemical system and surprisingly, this is not the case. Several conclusions can be drawn from this surprising revelation. The one that immediately comes to mind is that our biosphere and theirs share a common ancestry. They're eukaryotes, which means their cells have nuclei containing genetic material. Which suggests that their biosphere would have been separated from ours sometime after the appearance of this type of organism. The term Exo-Biospheric-Organism is actually a misnomer, but as it's a historical term, it's still used. Their genetics are not only based on the same genetic system, but they're also even compatible with our own cellular machinery. This means that you can take a human gene and insert it into an EBO cell, and that gene will be translated into protein, and this of course works in reverse with a human gene inserted into an EBO cell. There are important differences in post-translational modifications that will make the final protein non-functional, but I'll discuss these later. Their genome consists of 16 circular chromosomes.

You're probably familiar with the concept of intergenic region or "junk DNA". These are basically DNA sequences that don't code for proteins. These are evolutionary residues, transposons, inactivated genes and so on. To give you an idea, in humans, intergenic regions represent approximately 99% of our genome. I'm aware that these sequences aren't completely useless, they can be used as histone anchors, as buffers to protect coding DNA from radiation or even as alternative open reading frames, but that's rather peripheral.

What's particularly striking about the EBO genome is the uniformity of these intergenic regions. We see the same sequences repeated everywhere, and the distance in bp between the genes is virtually the same throughout their genome. The result is a minimalist, highly condensed genome. In fact, it's much smaller than ours. Moreover, the quantity of protein-coding genes is even significantly lower than ours, probably due to genetic refinement but also to biological processes that are absent in EBO. The uniformity of these sequences is a major indication of the artificiality of these beings. There is no complex organism on earth that has such elegance in its sequences. There is no evolutionary pressure that can lead to this kind of characteristic other than genetic engineering.

Speaking of genetic engineering, following sequencing of their genomes, we noticed a troubling and universal characteristic in the 5' of the regulatory sequence of each gene which we call the Tri-Palindromic Region. The TPR are 134bp sequences containing, as its name suggests, 3 palindromes. In genetics, a palindrome is a DNA sequence that when read in the same direction, gives the same sequence on both DNA strands. They serve both as a flag and as a binding site for proteins. The three palindromes in the TPR are distinct from one another and have been poetically named "5'P TPR", "M TPR" and "3' TPR". The TPR is composed (in 5' - 3' order) of 5'P TPR, 12bp spacer, Chromosomal address, 12bp spacer, M TPR, 12bp spacer, Gene address, 12pb spacer and 3' TPR. The chromosomal address is composed of 4 bp and is identical in each TPR of the same chromosome, but distinct between each of the 16 chromosomes of the genome. The Gene address is a 64bp sequence that is unique for each gene in the whole genome. It's therefore understandable that the TPR serves as a unique address not only for numerically identifying a gene, but also for identifying its chromosomal location. For those with only a basic knowledge of genetics, this is completely unheard of. No living thing in our biosphere has this kind of precise address in its genome. Once again, the presence of TPR cannot be explained by evolutionary pressure but only by genetic engineering on a genomic scale.

TPR opens the door to several possibilities. One of them suggests that EBO geneticists can insert or remove a gene from a cell in a way that is far more targeted and efficient than our technology allows. No proteins have been identified in the EBO genome that interacts with TPR. Rather, we believe that these proteins are exclusively targeted by external genetic engineering tools, probably used at the zygotic stage of embryonic development. The nature of these tools is unclear, but we definitely don't have anything like them. The probable absence of these proteins from the genome is a further indication of their artificiality. Given the high probability of artificiality of their genome and the apparent ease of modifying it with biomolecular tools, it's not out of the question that there could be polymorphism between individuals depending on their role and function. In other words, an individual could be genetically designed to have characteristics that give it an advantage in performing a given task, like soldier ants and worker ants in an anthill. Note that these previous statements are speculation. To my knowledge only one individual genome has been sequenced, I can't make a definitive statement on genetic variation between individuals.

I've talked a lot about intergenic regions, now I'll briefly discuss intragenic sequences. Briefly, because there's not a lot less to say despite its obvious importance. Much like ours, their genes have silencers, enhancers, promoters, 5'UTRs, exons, introns, 3' UTRs etc. There are many genes analogous to ours, which is not surprising given the compatibility of our cellular machinery. What's disturbing is that some genes correspond directly, nucleotide by nucleotide, with known human genes or even some animal genes. For these genes, there doesn't seem to be any artificial refinement but rather a crude copying and pasting. Why they do it is nebulous and still subject to conjecture. There are also many genes which are not found in our biosphere whose role has not been identified. Finding the purpose of these novel genes is one of the aims of the program. I'd like to note before going any further that this heterogeneity of genes of known and unknown origin is an undeniable proof of the artificiality of EBOs.

To conclude with genetics, the mitochondrial genome, at the time I was working there, had not yet been sequenced. It's safe to assume that this genome would also be streamlined and possibly has some version of TPR.

Transcription and translation and protein expression.

I briefly introduced the differences in post-translational modifications between human and EBO. This is hardly a surprise, as we often see the same thing between different terrestrial species. Obtaining a viable protein from a DNA sequence is a complex process involving hundreds of protein intermediates, each with a precise and essential role. A minor variation in this assembly line can lead to functional irregularities in the final product. So, it's no surprise that there are setbacks along the way when the first EBO gene transfection attempts failed to produce the desired functional protein in human cell lines. Fortunately for us, the work of what I imagine to be another team at another site has led to the development of an EBO cell line named EPI-G11 derived from epithelial tissues. With this tool in our hands, we were able to transfect and overexpress proteins of interest in order to eventually purify and study them. For your information, we use a biological ballistics delivery system (AKA gene gun) for our transfection needs because other methods are not very effective with cells of this line. For example, the viral vectors tested cannot be internalized by EPI-G11 and lipofection is too lethal. EPI-G11, like most eukaryotic cell lines, enters a phase of exponential growth when exposed to Fetal Bovine Serum. It's only half surprising that a cell line from such an exotic source should be sensitive to the growth factors present in FBS. In my opinion, this can be explained by the addition of animal genes to the genome, such as growth receptors.

Gross anatomy:

They are morphologically very similar to the grey aliens that are part of modern folklore. Their height is about 150cm, they have two arms, two legs and a head. Still, there are some notable differences.

Skin: The grey skin that is often described in folklore is in fact a biosynthetic film which, likely, serves to protect the EBO from a hostile environment. It doesn't provide effective protection against temperature changes, but it does offer adequate protection against the passage of liquids. It's possible that this film confers other advantages but my knowledge on the subject is limited. Under the grey film, the epidermis is rather white, and the texture is very regular and without any hair. We do not see any defect other than the folds near the joints. It's described as greasy in one report, but that's not something I've observed. The same report states that a strong, lingering smell of burnt hair and ammonia is present when the film is removed. There are a lot of pores on the skin, crossing from the epidermis to a gland in the hypodermis. These glands and pores are the terminal part of the excretory-sudoriferous system, which could explain the previously mentioned smell.

Head: The head contains two large, oversized eyes, two nostrils without protuberance, a narrow mouth without lips and two ear canals without auricles. There is a mandible, but the musculature is vestigial. There are no teeth or tongue in the oral cavity. The nasal cavity where the nostrils meet is compact and does not rise cranially but extends axially. There appears to be no equivalent to the olfactory bulb in the nasal cavity. The mouth leads directly to the esophagus and the nasal cavity to the trachea. The trachea and esophagus do not communicate.

Eye: Like the skin, the eyes are covered with a semi-transparent biosynthetic film that offers the same environmental protection, while providing protection against certain wavelengths and light intensity. When the film is removed, a more traditional eye is revealed. It's about three times larger than a human eye and there are no eyelids. The size of their eyes suggests they have excellent night vision. It seems paradoxical to cover them with a semi-opaque film. Perhaps they only need to wear it in a bright environment. Their sclera is the same color as their skin, the iris is pale grey, and the pupil is black and oversized. The lens is rounder than a human, and the musculature used to adjust focus is more developed. On the retina, there are at least 6 types of cone cells. The responsiveness of each of these 6 types of cone is specific to a wavelength band, with a minimum of overlap between each other. The result is a broader visible spectrum.

Ear: As mentioned, the outer ear has no auricle and the ear canal is unremarkable. The inner ear has all the characteristics of a typical vestibular and cochlear system, although the curvature of the cochlea is more pronounced than a human. This probably results in greater hearing acuity for low frequencies.

Brain: The brain is tetraspheric, i.e. composed of four major sections. The sections are separated by transverse and longitudinal fissures and are connected to the central lobe, which acts as brainstem and cerebellum. The volume of the brain is around 20% superior to that of a man of the same height. It has a much more pronounced level of gyrification than an average human. Moreover, the ratio of glial cells to neurons is also slightly higher than in humans. It is important to mention the presence of nodules on the central lobe. Histological analysis of these structures reveals a kind of intricate biological circuitry. It is speculated that these nodules are essential to interact with their technology. Consequently, determining the proteome of these structures is an absolute priority for the program.

Neck: The neck is proportionally longer than that of a human, and at the same time relatively thin. As mentioned, the esophagus and trachea are separate. There are no vocal cords in this region.

Thorax: The musculature of the thorax is underdeveloped. Muscles equivalent to the pectoralis major can be seen. We can also see the trapezius and deltoid muscles. The sternocleidomastoids are well defined. The ribs and sternum are clearly visible. There are no nipples.

Abdomen: The abdomen is wider than the thorax and bulges slightly forward. There is no navel.

Pelvis: The pelvic bones are apparent. There are no genitals or anus.

Hands and feet: Their hands have four digits, including an opposable thumb on the medial side. They have no nails, and the texture of their fingerprints is composed of concentric circles. Fingers are proportionally much longer than in humans. Unlike humans, finger musculature is entirely intrinsic to the hand. In other words, the muscles used to move the fingers are not in the forearms but entirely located in the hands. At first glance, the feet consist of just two digits, but a necropsy soon determined that each toe was made of two fused digits. The medial toe is marginally longer than the distal toe. The feet are relatively longer and narrower than in a human. Their musculature, however, is vestigial.

The EBOs endoskeleton is very similar to ours, at least in terms of composition. There's collagen, hydroxyapatite but also copper oxide crystals where marrow would normally be found. The role of these crystals has not been established, but it is not a crystalopathic condition. The blood cells of the myeloid lineage (or the equivalent for these creatures) therefore mature in a different location than in humans i.e. in the thymus-like organ. A transverse section of the bone reveals osteon and osteocytes. There appear to be few osteoblasts and no osteoclasts. This indicates that the bones are no longer growing and cannot absorb the minerals present or adapt mechanically to changes in posture.

Biological system:

Respiratory system: Their cellular respiration is equivalent to ours, i.e. they need to oxidize organic components to produce energy. Their lungs have no reciprocating action, but rather have a unidirectional flow of air, similar to those seen in birds, which is more efficient than ours. It is speculated that this is in response to the brain's elevated metabolic needs. Vocalization is produced by vibration of the wall membrane at the junction between the two air sacs.

The Circulatory system of EBOs is rather analogous to ours. The heart is located in the mediastinum, but in a more medial position, directly beneath the sternum. The heart has two ventricles and two atria. There is an aorta, a pulmonary vein, a pulmonary artery and a vena cava. Blood flowing to the pulmonary capillaries via the pulmonary artery is pumped against the flow of air, maximizing gas exchange efficiency. The blood gas barrier is relatively narrow in these capillaries, at least compared to a human. Then oxygen-rich blood is returned to the heart and then expelled into the aorta and the rest of the body. Before returning to the heart, the blood will pass through the hepatorenal organ which, among other things, filters and controls osmotic pressure of the blood.

The blood itself is also analogous to that of a human. However, the proportion of plasma is much higher, albumin is in similar proportion, hormone levels are much lower, metal ion levels are much higher (particularly copper) and glucose levels are significantly higher. The color of the blood is brownish, given the higher proportion of plasma and concentration of metal ions. On the cellular side, there are erythrocytes which, in addition to hemoglobin for binding oxygen, display several complexes capable of binding copper ions. It's not clear what role these copper ions play but we believe it neutralizes blood ammonia, among other things. Several cell types with leukocyte characteristics have been observed, but no comprehensive knowledge of them exists. Platelets are present, but in smaller proportions than in humans.

Excreto-sudoriferous system: This system is completely different from what I've seen. As mentioned earlier, there is no large orifice, like an anus or urethra, to get rid of biological waste. Instead, there are countless small pores on the surface of the skin. There's a large medial organ called the hepatorenal organ, which acts as both kidney and liver and is central to maintaining homeostasis. This organ is highly vascularized and the blood must pass through it before returning to the heart. Its role is, among other things, to purify the blood of metabolic waste. Waste is excreted into the equivalent of a ureter, which branches out into four. Each branch flows towards one of the four limbs and in turn these branches divide until they end up as thousands of excretory pores. The motility of this excretory system is mediated by a weak peristalsis at the proximal level and on the four main branches. Peristalsis ceases around the first distal junction. As there is no urea cycle, the ammonia concentration at the exit of the hepatorenal organ is very high. This ammonia is carried to the pores and gives the distinct odor I mentioned earlier. The rationale behind this unusual excretory system is directly related to this excreted ammonia, which enables thermoregulation by evaporating on the skin's surface. The greater the physical effort, the greater the metabolism. This in turn leads to a rise in temperature, and a corresponding increase in metabolic waste via amino acid catabolism. This leads to an increase in filtration and ammonia excretion, which ultimately lowers body temperature.

Digestive system: The digestive system is extremely underdeveloped. There's no stomach in the familiar sense. However, there is a pseudo-stomach located at the transition between the thoracic and abdominal cavities. This organ is not involved in digestion, but only serves as a reservoir. A sphincter controls the flow of food into the intestine. The intestine is limited to the equivalent of our small intestine, i.e. it only serves to absorb liquids and nutrients and acts as the main digestion site. It has villi and microvilli like ours. The intestine ends in the hepato-renal organ, where non-digested matter is transported to the ureter and excretory system. Residues are dissolved in the ammonia of metabolic waste for excretion. There's an organ near the pseudostomachal sphincter that secretes digestive enzymes directly into the intestine. This organ is inspirationally called the digestive organ. It secretes mainly proteolytic enzymes and glycoside hydrolases.

Given the absence of teeth, the narrowness and rigidity of the esophagus, the absence of a true stomach and the absence of defecation, it is strongly believed that EBOs can only consume food in liquid form. It is assumed that, given the high metabolic needs of their brains, this food would have a high carbohydrate concentration. In order to meet other metabolic needs, there must also be a high protein content in the food consumed. These two statements are supported by the type of enzyme secreted by the digestive organ. It is therefore speculated that the food consumed is a sort of broth rich in sugar and protein, which probably also has a high copper content. Given the strict limitations on the type of food that they can consume, it's unlikely that this type of creature could survive in our biosphere without technological support.

Endocrine system: Knowledge of the endocrine system is minimal. We know that cells are receptive to bovine growth hormones, so it's assumed that certain functions are regulated by such a system. Endocrine mechanisms are very complex, and it goes without saying that they are best studied on living subjects.

Immune system: The immune system is another unknown. There seems to be an innate immune system but there doesn't seem to be any adaptive immunity, at least not similar to what is known. There's a thymus-like organ near the heart that's proportionally larger than in humans. This organ seems to be where all blood cells mature. Some cells have leukocyte characteristics such as granularity. The immune cells that germinate here have a high copper concentration. The surface receptors of innate immune cells have not yet been characterized, so we might as well say that all the work remains to be done.

Nervous system: The nervous system is also relatively similar. The spinal cord begins at the base of the central lobe of the brain and propagates down the vertebral column. In the vertebrae there are ganglia made of afferent and efferent neurons. In short, other than the CNS, there is nothing out of the ordinary.

Musculoskeletal system: The musculoskeletal system is very ordinary, albeit underdeveloped. Most of the human skeletal muscles have an equivalent. Only the hands, feet and forearms are different. It should be noted that the proportion of type 1 and type 2 muscle fibers is different from that in a human. Indeed, type 1 outnumbers type 2 by about a factor of 10.

Artificial system: We speculate that artificial molecular machines may be present in the body, and that copper, if present, would be essential to their function or assembly. Importantly, no AMMs have been observed

Question 1: Amazing story. Have you shared this with the Senate Select Commission on Intelligence or with AARO and do you have evidence to back this up?

Thank you, no I haven't and no I won't. It sounds like a honey trap to me. I will not place my life in the hands of politicians. I have no proof other than this message. I know it's not much but it's what I'm prepared to offer

Question 2: Well that was a read ... So they are bio engineered worker bees... Any elemental components that are unattributal to our biome?

Yes, knowing that they're disposable, unable to live independently without technological support, and that they're ephemeral. The only suitable hypothesis is that they are alive only to accomplish their task. Can you clarify your question about elemental components?

Question 3: I haven't read everything in detail but can you expend on the document on their religion?

EBOs believe that the soul is not an extension of the individual, but rather a fundamental characteristic of nature that expresses itself as a field, not unlike gravity. In the presence of life, this field acquires complexity, resulting in negative entropy if that makes sense. This gain in complexity is directly correlated with the concentration of living organisms in a given location. With time, and with the right conditions, life in turn becomes more complex until the appearance of sentient life. After reaching this threshold, the field begins to express itself through these sentient beings, forming what we call the soul. Through their life experiences, sentient beings will in turn influence the field in a sort of positive feedback loop. This in turn further accelerates the complexity of the field. Eventually, when the field reaches a "critical mass", there will be a sort of apotheosis. It's not clear what this means in practical terms, but this quest for apotheosis seems to be the EBOs main motivation.

The author of the document added his reflections and interpretations as an appendix. He specified that, for them, the soul field is not a belief but an obvious truth. He also argues that the soul loses its individuality after death, but that memory and experience persist as part of the field. This fact would influence the philosophy and culture of EBOs, resulting in a society that doesn't fear death but which places no importance or reverence on individuality. This 'belief' compels them to seed life, shape it, nurture it, monitor it and influence it for the ultimate purpose of creating this apotheosis. Paradoxically, they have little or no respect for an individual's well-being.

Please be advised that I'm speaking from memory of something I read more than 10 years ago, so take the following with a grain of salt. Also, I'm not a philosopher or an artist, so please excuse my struggle to properly formulate the concepts and my dry terminology. Finally, note that this information comes from a document whose author was directly interacting with an EBO. It is not specified whether it was an ambassador, a crash survivor, a prisoner. The means of communication were not specified either.

BIBLIOGRAPHY

Banduric, Richard B. 2018. Breakthrough Spacecraft Propulsion Concept by Breaking Green's Reciprocity. Submitted to the 9th JANNAF [Joint Army-Navy-NASA-Air Force Interagency Propulsion Committee] Spacecraft Propulsion Conference, advance propulsion session. Featured at https://www.linkedin.com/in/richard-banduric-53232667/.

Bartlett, Frederic C. 1932. *Remembering: A Study in Experimental and Social Psychology*. Cambridge University Press. Reprint (2nd ed.): 1995.

Bradshaw, Corey J. A., Christian Reepmeyer, Theodora Moutsiou, et al. 2024. *Demographic models predict end-Pleistocene arrival and rapid expansion of pre-agropastoralist humans in Cyprus. PNAS* 121 (21):e2318293121.

Cameron, Grant, and Desta Barnabe. 2022. *UFO Sky Pilots: Pilots of Peace and Oneness*. Itsallconnected Publishing.

Carlotto, Mark C. 2018. *Before Atlantis: New Evidence of a Previous Technological Civilization*. CreateSpace.

Coppo, Alessandro, Alessandro Cuccoli, and Paola Verrucchi. 2024. Magnetic clock for a harmonic oscillator. *Physical Review A* 109:052212.

Corso, Phillip J., Col., with William J. Birnes. 1997. *The Day After Roswell*. Pocket Books. Reprint: 2017. Gallery Books/Simon & Schuster.

Dennett, Preston. 2006. *Human Levitation: A True History and How-To Manual*. REDFeather/Schiffer Publishing.

Edwin A. Abbott. 1884. *Flatland: A Romance of Many Dimensions*. Seeley & Co. (Many reprints)

Eire, Carlos M. N. 2023. *They Flew: A History of the Impossible.* Yale University Press.

Ellis, John, Brian D. Fields, and David N. Schramm. 1996. Geological Isotope Anomalies as Signatures of Nearby Supernovae. *The Astrophysical Journal* 470:1227–1236. https://adsabs.harvard.edu/full/1996ApJ...470.1227E.

Eno, Paul. 2019. *Dancing Past the Graveyard: Poltergeists, Parasites, Parallel Worlds, and God.* REDFeather/Schiffer Publishing.

Freyd, Jennifer J. 1996. *Betrayal Trauma: The Logic of Forgetting Childhood Abuse.* Harvard University Press.

Grandin, Temple. 1995. *Thinking in Pictures: My Life with Autism.* Doubleday. Expanded 25th anniversary edition: 2008. Vintage.

Gurdjieff, G. I. 1950. *All and Everything: Beelzebub's Tales to His Grandson.* Harcourt Brace. Reprint: 1999. Penguin Compass.

Hancock, Graham. 2002. *Underworld: The Mysterious Origins of Civilization.* Crown.

Huels, E. R., H. Kim, U. Lee, T. Bel-Bahar, A. V. Colmenero, et al. 2021. Neural correlates of the shamanic state of consciousness. Frontiers in Human Neuroscience 15, 140. https://doi.org/10.3389/fnhum.2021.610466.

Kozhevnikov, M., J. Elliott, J. Shephard, and K. Gramann. 2013. Neurocognitive and somatic components of temperature increases during Tummo meditation: Legend and reality. *PLoS One* 8 (3): e58244.

Kuiper, T. B. H., and M. Morris. 1977. Searching for extraterrestrial civilizations. *Science* 196:616–621.

Leir, Roger. 1998. *The Aliens and the Scalpel: Scientific Proof of Extraterrestrial Implants in Humans.* Granite Pub. Revised 2nd ed.: 2005. The Book Tree.

Leir, Roger. 2005. *UFO Crash in Brazil: A Genuine UFO Crash with Surviving ETs.* The Book Tree.

McGilchrist, Iain. 2009. *The Master and His Emissary: The Divided Brain and the Making of the Western World.* Yale University Press.

Morrow, Susan Brind. 2015. *The Dawning Moon of the Mind: Unlocking the Pyramid Texts.* Farrar, Straus and Giroux.

Norris, Ray P., and Barnaby R. M. Norris. 2020. "Why are there Seven Sisters?" arXiv2101.09170. https://arxiv.org/pdf/2101.09170.

Pasulka, Diana Walsh. 2019. *American Cosmic: UFOs, Religion, Technology.* Oxford University Press.

Powell, Diane Hennacy. 2008. *The ESP Enigma: The Scientific Case for Psychic Phenomena.* Walker & Company.

Pratt, Bob. 1996. *UFO Danger Zone: Terror and Death in Brazil—Where Next?* Horus House.

Prince-Hughes, Dawn. 2004. *Songs of the Gorilla Nation: My Journey Through Autism.* Harmony. Reprint: 2005. Crown.

Randle, Kevin D., and Donald R. Schmitt. 1991. *UFO Crash at Roswell.* Avon.

Rodney, Charles, and Anna Jordan. 1995. *Lighter than Air: Miracles of Human Fight from Christian Saints to Native American Spirits.* 1st World Publishing.

Saebels, Corina. 2007. *The Collectors: A Canadian UFO Experience.* Trafford Publishing.

Strieber, Anne, and Whitley Strieber. 1997. *The Communion Letters.* Harper Prism.

Strieber, Whitley. 1987. *Communion.* Beech Tree.

———. 1995. *Breakthrough: The Next Step.* HarperCollins.

———. 2012. *Solving the Communion Enigma: What Is to Come.* Tarcher.

———. 2020. *A New World.* Beyond Words.

———. 2023. *Them.* Walker & Collier.

Strieber, Whitley, and Jeffrey J. Kripal. *The Super Natural: Why the Unexplained Is Real.* 2016. Walker & Collier. Reprint: 2017. TarcherPerigee.

Tart, Charles. 1968. A psychophysiological study of out-of-the-body experiences in a selected subject. *Journal of the American Society for Psychical Research* 62:3–27.

Tegmark, Max. 2014. *Our Mathematical Universe: My Quest for the Ultimate Nature of Reality.* Alfred A. Knopf.

Vallée, Jacques. 2014. *Passport to Magonia: From Folklore to Flying Saucers.* Daily Grail Publishing.

von Neumann, John. 1966. *Theory of Self-Reproducing Automata.* University of Illinois Press.

Wahbeh, Helané, Dean Radin, Garret Yount, et al. 2022. Genetics of psychic ability – A pilot case-control exome sequencing study. *Explore* 18 (3):264–271.

www.ingramcontent.com/pod-product-compliance
Lightning Source LLC
Chambersburg PA
CBHW060452030426
42337CB00015B/1564